200 7/24

A Pictorial History of Robins Air Force Base, Georgia

A PROJECT OF THE
ROBINS AIR FORCE BASE
HERITAGE COMMITTEE

1982

AIR FORCE LOGISTICS COMMAND

Dedication

This volume is dedicated to the men and women of Robins Air Force Base past and present who, while providing security for the United States of America, have created a proud heritage to guide the future men and women of Robins Air Force Base, Georgia, and the nation.

Foreword

America today is in the first years of its third century as a free nation. That freedom was won by men and women who did what they had to do, and did it well. Their descendants are still free. The history of the United States armed forces is the story of a struggle that began in the eighteenth century. As we approach the twenty-first century, the faces of our enemies have changed, but danger to our freedom is still with us. As part of those armed forces, we can never forget the lessons of the past.

In September 1981, a charter to preserve the past of the Air Force Logistics Command was established by our Commander, General James P. Mullins. As always, the response from our people has been outstanding. Under the guidance of senior civilian Earvin L. "Pete" Joyner, a Robins Air Force Base Heritage Committee has put into action a program which will highlight a proud record we share with the people who have always given their best to this command.

Memorialization of base streets and buildings and organizational unit displays will honor our workforce and their past and present contributions to national defense. Phase one of a privately-funded museum of aviation has been approved by the Air Force for establishment on base land. The museum will receive minimal support from the Air Force, but the interest and funds needed to make it a fine museum must come from the same base-community partnership which led to the construction of Robins Field at Wellston, Georgia in 1941. This volume of history is presented as a lasting record of events which go back to that beginning.

Robins Air Force Base was born in a decade that brought the major war of our time and with it, a pressing need for logistics support to the air defense forces of the United States. That need did not end with World War II, for air power had come into its own. The world entered an era of uneasy peace broken by conflicts that seemed more like wars, and Robins Air Force Base moved through four decades of challenge and tremendous growth. As you read these pages, the Air Force Logistics Command is at work making the history of tomorrow. Because our logistics world never stands still. As fast as defense systems change, we must change. Those changes will continue to have a major impact on our citizen partners in defense.

The friendship and mutual support between Robins Air Force Base and its surrounding communities go back a long way. From the initial donation of land which, with the help of distinguished Georgia Congressman Carl Vinson, brought a major air depot to Middle Georgia, our friendship has stayed on course. That story is a memorable chapter in the history of Robins Air Force Base.

Each part of this forty-year story took many months of research and just plain hard work by Lew Brinson and Carolyn Petty leading the Pictorial History writers: Herb Stephens, Joe Molony, Cara Best and Dee Teal. It could not have been written without the enthusiastic support of those citizens and airmen who have taken part in the dynamic progress of Robins Air Force Base over the past four decades. To each, we owe a debt of gratitude.

The Pictorial History of Robins Air Force Base is a remembrance of things past. But it will also stand as a tribute to the "can do" spirit shown by the people of Robins Air Force Base and Middle Georgia. That spirit, multiplied many times throughout this command, at bases and communities across the country, is the solid foundation of a free future for all Americans. General Mullins has said, "Historically, the great innovators were people with the vision and courage to do the impossible." Read this history as a reminder that the men and women of the Air Force Logistics Command have done, and will continue to do the impossible. That is our real heritage.

John R. Paulk
Major General, Commander
Warner Robins Air Logistics Center

ISBN 0-88146-002-8

Produced by

UNIVERSITY PRESS OF THE SOUTH
1400 Coleman Avenue
Macon, Georgia 31207

Composition by
OMNI COMPOSITION SERVICES
Macon, Georgia 31207

Designed by
HAYWOOD L. ELLIS

ISBN 0-88146-002-8

Library of Congress Cataloging in Publication
Number 82-20197

Table of Contents

Left to right—Mr. Joyner, General Mullins and General Paulk.

Acknowledgements

It is with great pride that the Robins Air Force Base Heritage Committee presents to you this book on a tradition of service to our country by more than four decades of Americans who have called Robins Air Force Base home.

First, we acknowledge the initiative of General James P. Mullins, AFLC Commander, who sought a history from each of the Air Force Logistics Command bases. Aware that the Traditions of excellence which have characterized the United States Air Force and its Logistics Command are worthy guides to future generations, General Mullins in one of his early decisions directed that each installation develop a Heritage Program, and that the Heritage Program include a pictorial history. His guidelines provided directions for our people in preparing this volume.

Second, we acknowledge the leadership provided by Major General John R. Paulk, among whose early actions in 1980 were the establishment of historical displays at key locations on base, getting our Heritage Program off to an advanced start. The Heritage Committee expresses its thanks to General Paulk for allowing it adequate resources to create this Pictorial History.

Third, our thanks go the members of the special history cadre, who are identified on the last page of this book.

Fourth, we appreciate the involvement of Robins people, present and past, in the enthusiastic furnishing of photographs and memorabilia and most particularly to the following people who joined me in spending hours in editing the text: A. H. Cotton, whose career at the apex of Warner Robins Leadership will be difficult to replicate in length of service and in quality of Leadership; Jack H. Braucht, whose career at Robins spanned most of the decades of its history, and whose perspective on the non-Air Corps history during the 1940s was particularly significant; L. M. "Pete" Rucker. whose counsel from the Command Section guided the cadre effectively; Dr. Richard W. Iobst, WR-ALC Historian and Robert J. Smith, Command Historian, who provided historical scholarship to the writers of the book in gentle ways; to Peggy Young, who aided in the selection and captioned the color photographs in Chapter 5; to Barbara Pirkle of the History Office whose research was timely, accurate and cheerful, even on the thousandth request; to Colonel Forrest E. Storz, Robert L. Burns, and Theodore M. Demarest of the Directorate of Plans and Programs, who provided agile minds and full support to the cadre during the writing; to the men and women of the Plans and Programs Division—Barry Campbell, Sue Crites, and Randy Lanier, among them—who sustained the functions of the division while three of its members were on this special assignment and to Robbie Crowder and Addie Padgett who supported the text preparation; and to Dr. Alain Posteau of the History staff for his careful attention to detail. And to those persons listed in the appendix who opened their personal archives to the team, our sincere thanks.

Fifth, the team who did the printing for us has done a handsome job. To Earl Buckner of the Administration Division, Robert McDonald, Jim Duckworth and Lt. Col. E. Randolph Silsbee (USAF, Ret.) of the Audiovisual Branch, and to Dr. Watson Mills and Harold Butts and the able staff of the Mercer University Press and University Press of the South go our sincere congratulations.

Earvin L. Joyner
Deputy Director of Maintenance
Chairman, Robins Air Force Base Heritage Committee

Prologue

From the tenth of May until the twenty-second of June 1940, the world was witness to the most astonishing military campaign of the twentieth century. In a mere forty-four days the German Army occupied Luxemburg without resistance, defeated the Netherlands, Belgium and France and compelled the evacuation of the British Expeditionary Force from the French seaport of Dunkirk. If that sounds bad, it was actually worse. The campaign had been lost by nightfall of the fifth day. They called it "Blitzkrieg"—"lightning war."

On this side of the Atlantic, a people who had talked of "splendid isolation" and "barrier oceans" began to feel more than uneasy. The German conquest of Poland in September 1939 had not caused such consternation. The Polish Army had been considered primitive by western standards. The British and French, on the other hand, had been thought to be more than a match for the Germans. Thus the swift fall of France was the greatest military shock that the American public had suffered since the Civil War. When Maj. Gen. Henry H. Arnold, Deputy Chief of Staff for Air, went before Congress that June, he was told that any appropriation he needed would be forthcoming. "All you have to do is ask for it," said Senator Henry Cabot Lodge, Jr. The feverish rearmament which proceeded can be seen from the statistics. In 1939 there were twenty-one Army air installations in the continental United States. By the end of 1941, there were 114 completed or under construction—including the one which is now known as Robins Air Force Base. Little did anyone realize at the time, but the United States of America was setting out upon a path from which, to this day, it has never been able to turn back.

CHAPTER ONE: *The 1940s: A Foundation Laid*

Houston County

Houston County, the site of Robins Air Force Base, was not named for Sam Houston. It was named for John Houstoun and is pronounced "House-ton." John Houstoun belonged to an English family of some prominence but cast his lot with the colonists in the American Revolution. He was the second Governor of Georgia (1778). Just how and when the second "u" in the name dropped out is unknown. What is known is that earlier generations were not so particular about proper spelling. John Houstoun never saw the county which bears his name.

TREATY WITH THE CREEKS, 1821.

Articles of a treaty entered into at the Indian Spring, in the Creek Nation, by Daniel M. Forney, of the State of North Carolina, and David Meriwether, of the State of Georgia, specially appointed for that purpose, on the part of the United States; and the Chiefs, Head Men, and Warriors, of the Creek Nation, in council assembled. Jan. 8, 1821. 7 Stat., 215. Proclamation, Mar. 2, 1821.

ART. 1. The Chiefs, Head Men, and Warriors, of the Creek Nation, in behalf of the said nation, do, by these presents, cede to the United States all that tract or parcel of land, situate, lying, and being, east of the following bounds and limits, viz: Beginning on the east bank of Flint river, where Jackson's line crosses, running thence, up the eastern bank of the same, along the water's edge, to the head of the principal western branch; from thence, the nearest and a direct line, to the Chatahooche river, up the eastern bank of the said river, along the water's edge, to the shallow Ford, where the present boundary line between the state of Georgia and the Creek nation touches the said river: Cession by the Creek. Boundaries.

Done at the Indian Spring, this eighth day of January, A. D. eighteen hundred and twenty-one.

D. M. Forney,	[L. S.]	Taskagee Emauthlau, his x mark,	[L. S.]
D. Meriwether,	[L. S.]	Tuckle Luslee, his x mark,	[L. S.]
Wm. McIntosh,	[L. S.]	Tuckte Lustee Haujo, his x mark,	[L. S.]
Tustunnugee Hopoie, his x mark,	[L. S.]	Cunepee Emauthlau, his x mark,	[L. S.]
Efau Emauthlau, his x mark,	[L. S.]	Hethlepoie, his x mark,	[L. S.]
Holoughlan, or Col. Blue, his x mark,	[L. S.]	Tuskeenaheocki, his x mark,	[L. S.]
Cussetau Micco, his x mark,	[L. S.]	Chaughle Micco, his x mark,	[L. S.]
Sotetan Haujo, his x mark,	[L. S.]	Isfaune Tustunnuggee Haujo, his x mark,	[L. S.]
Etomme Tustunnuggee, his x mark,	[L. S.]	Wau Thlucco Haujo, his x mark,	[L. S.]
Itchu Haujo, his x mark,	[L. S.]	Houpauthlee Tustunnuggee, his x mark,	[L. S.]
Alabama Tustunnuggee, his x mark,	[L. S.]	Nenehaumaughtoochie, his x mark,	[L. S.]
Holoughlan Tustunnuggee, his x mark,	[L. S.]	Henelau Tixico, his x mark,	[L. S.]
Auhauluck Yohola, his x mark,	[L. S.]	Tusekeagh Haujo, his x mark,	[L. S.]
Oseachee Tustunnuggee, his x mark,	[L. S.]	Joseph Marshall,	[L. S.]

In presence of—

I. McIntosh,
David Adams,
Daniel Newman,
Commissioners of Georgia.
D. B. Mitchell, Agent for I. A.

William Cook, secretary C. G.
William Hambly,
Sl. Hawkins,
George Levett,
Interpreters.

Excerpts from the Creek Treaty of 1821 from the journal of the Georgia House of Representatives. Of the twenty-six Indian signatories, two had non-Indian names and these two were the only ones who could actually write their names—Joseph Marshall and William McIntosh.

Houston County was created from land ceded to the State of Georgia by the Creek Indians in the Creek Treaty of 1821. The treaty was negotiated at Indian Springs, which is now a state park in Butts County, above Macon. The Upper Creeks rejected the treaty but a delegation of Lower Creeks, led by the controversial half-white Chief, William McIntosh, signed it. McIntosh was not only a Creek Chief but a Brigadier General, USA, first cousin of Governor George M. Troup, and a rich man with numerous slaves. He had three wives, two red and one white, but attended the Methodist Church when he visited his cousin, the Governor, in Milledgeville. Milledgeville was the state capital in those days. A second Creek Treaty (1825) negotiated at Indian Springs, in which McIntosh ceded another slice of Indian land, cost him his life. He was shot and scalped by a party of Creek warriors. It didn't change anything. The remaining Indians of the Southeastern United States were rounded up in the 1830s and moved to what is now the state of Oklahoma. This accounts for the similarity of Indian place names, e.g., Oakmulgee (city), Oklahoma and Ocmulgee (river), Georgia.

Nothing of particular historical interest happened within the present borders of Houston County. No Revolutionary War battles were fought here and the nearest approach of Gen. William T. Sherman's Army was the northeastern outskirts of the City of Macon. Samuel Rumph's plantation was originally in Houston but is now in Macon County as a result of one of the many partitions and border revisions which, by 1924, had reduced Houston to about one-third its original size. Rumph developed the Elberta Peach—the first delicious variety tough enough to withstand shipment to distant markets. He thus originated the commercial peach industry in the United States. That resulted in Georgia being named the "Peach State."

Houston County, 1847 Bonner Map. Wellborns Mills was the nearest town to what is now Robins Air Force Base.

The land which is now the site of Robins Air Force Base was assembled from numerous individual property owners but much of it was once a part of the Feagin Plantation. In fact, the grave of Henry Feagin (1795-1842) is on Robins Air Force Base in the tiny cemetery on 10th Street, in the block between E and F Streets. His home, "the Oaks," was also on Robins Air Force Base and was still in existence when the base was built. For a time, during World War II, it was used as the enlisted men's Guest House. In 1966 the house was sold and moved away from Robins AFB. Its current whereabouts are unknown.

The majority of the population of Houston County is composed of relative newcomers, i.e., they or their families became residents of the county after 1940. For them, the good old days began with the construction of Robins Air Force Base. The golden age is here and now.

Geography—It Mattered

If you were to draw a line across a map of Georgia, connecting the cities of Augusta, Macon and Columbus, you would have drawn the approximate shoreline of a prehistoric ocean. This line that you just drew is called the "Fall Line." The land above the line is called the Piedmont. It is rolling countryside, steep and broken in places, until it finally erupts into the Appalachian Mountains. The rivers of the Piedmont are not navigable, its soil is generally red clay with out-

Surveying, Mapping and Distributing the New Indian Lands

After the acquisition on January 8, 1821, of the Indian lands in the Creek Treaty of 1821, the Georgia General Assembly passed and Governor John Clark signed on May 15, 1821, an act establishing five new counties (Dooly, Houston, Monroe, Fayette, and Henry) and specifying how these lands were to be distributed to the citizens of the state. The borders of Houston were defined to be "All that part of said territory lying between a line commencing on the Ocmulgee river, opposite Fort Hawkins, and running due west to Flint river, and the line first above described [which described the bounds of Dooly County] shall form one other county, to be called Houston." The act further specified that each county should be laid out and described into districts of nine miles square ("as near as practicable"), and that the districts be subdivided into square tracts containing each two hundred and two and a half acres, marked and numbered, and that surveyors be named to survey each district under the supervision of the Surveyor General of the State of Georgia. These surveys, paid for at the rate of $3.50 per mile with no provision for incidental expenses, are on file in the Surveyor General Department in the State Archives Building in Atlanta.

This system of marking and numbering each Land Lot (the "square tract of two hundred two and a half acres" is called a "Land Lot") makes identification of the owners of the land which became Robins Air Force Base an easy task. The land lots were distributed in a lottery, established by the State. The basic qualification to participate in the lottery was "every male white person of eighteen years of age and upwards being a citizen of the United States at least three years, and an inhabitant to this State three years immediately preceding the passage of this act." Also allowed to participate were widows and orphans of veterans of "the late wars against Great Britain or the Indians."

Although the plan of distributing the new land by lottery was designed to encourage the lucky citizens to move to and develop it, the usual custom was the selling of the lots to speculators who then sold the land to those interested in settling on it. A list of the original owners of the land which became Robins Air Force Base is included as an appendix.

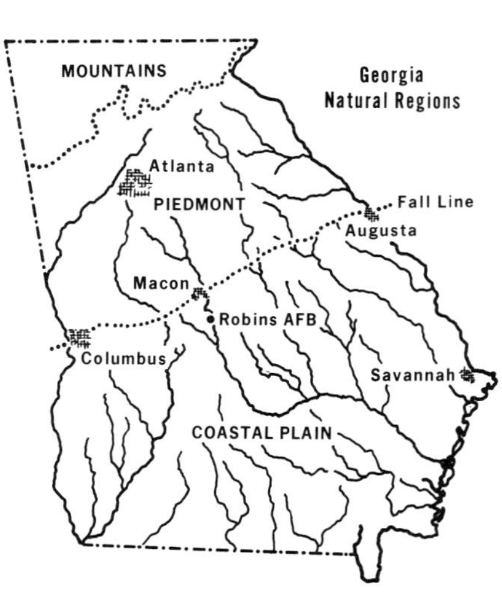

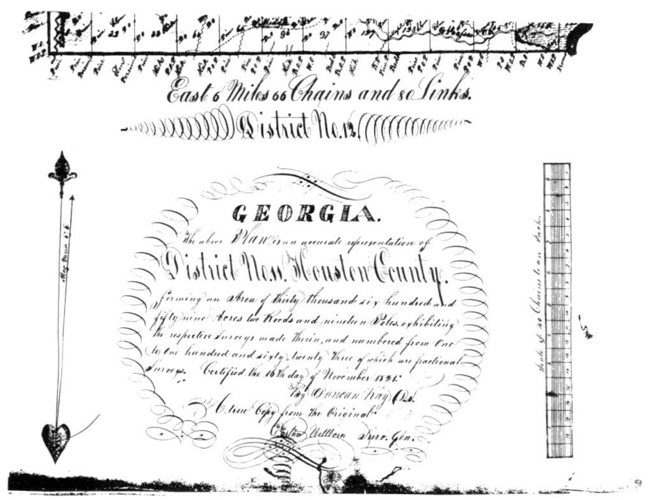

Document signed by Carlton Wellborn as Surveyor General of Georgia. It certified that the document was a true copy of the land survey of the Eleventh District of Houston County.

croppings of igneous and metamorphic rock, and its best agricultural uses are forest and pasture.

The land below the line is called the Coastal Plain. It is relatively flat and becomes even more so as it stretches toward the seashore. The rivers are navigable and the few rocks to be found are sedimentary. The clay washed down from the Piedmont is diluted with sand to make a loam and the area is favorable for general agriculture. The purest water is both shallow and abundant.

This description is, of course, not unique to Georgia. It is the common geography of the eastern United States between the Atlantic Ocean and the Appalachian Mountains. Neither is the location of Macon, Columbus and Augusta on rivers at the Fall Line unusual. Richmond, Virginia; Raleigh, North Carolina; and Columbia, South Carolina, to name a few, were built at such points and for the same reasons. The sudden change of topography creates rapids and waterfalls which can be used to power machinery. Also, these sites are at the head of navigable streams to the ocean, providing natural transportation terminals for converting from overland to river travel modes.

In the Old South, the Coastal Plain was the center of wealth, culture, political power, and slavery. In the New South, prosperity has largely shifted to the Piedmont. Such cities as Macon, on the border between the two regions, have been able to prosper in both eras.

First North Houston Place: Wellborns Mills

In Fussell M. Chalker's *Pioneer Days Along the Ocmulgee*, the pioneer life is described: "The grist mill was a favorite gathering place for the pioneer men. Collecting a supply of corn, they would set out on horseback or by ox cart for the neighborhood mill. Here, the men would find congenial companions and, while waiting for their corn to be ground, they would engage in lively discussions of daily happenings. The local grist mills frequently became the centers around which populous settlements arose."

In a deed of sale dated October 1, 1830, a sixty acre tract of land in the Eleventh District of Houston County is described as you would expect, but one corner point is of particular interest: "to a black Jack Bush near the road leading from Macon to Wellborns Mill."

"Wellborns Mills" is cited as being a post office in 1837 in the third edition of the *Gazetteer of the State of Georgia*, written and published by a noted Georgia Baptist preacher, Adiel Sherwood.

Who was Wellborn, and where was his mill?

Wellborns Mills is shown on the 1847 Bonner map of Georgia as being on the road from Macon to Hawkinsville at the crossing of Sandy Run Creek, the only "place" on that long road at that time.

The originator of Wellborns Mills was a distinguished early settler of the region, Carlton Wellborn, born in 1787, a native of Wilkes County. Mr. Wellborn had been appointed by the Legislature on November 24, 1825, to the post of Surveyor General of the State of Georgia for a two year period. Administering the surveys of the rapidly expanding state during these boom periods was a formidable task, demanding the presence of the Surveyor General in the capitol at Milledgeville. It was as a citizen of Baldwin County that Mr. Wellborn first invested in Houston County, buying Land Lot 81 of the Eleventh District for $218.50. It was upon this land lot that Mr. Wellborn would soon establish his mill, on the waters of Sandy Run Creek.[1]

Mr. Wellborn moved to and continued to invest in Houston County property in the 1830s. He was the wealthiest citizen of the area in the census of 1850, owning real estate valued at $19,500 as well as 34 slaves. He was active politically, representing Houston County in the State Legislature in 1830, and at state conventions in 1832 and 1833. His daughter Caroline married Barnett Holleman who farmed in the county. His son, Oliver, was also a Houston County farmer.

In 1973, the Base expanded to the south in an exchange of lands with the governments of Houston County and the City of Warner Robins. A portion of that land had been purchased from the great, great-granddaughter of Carlton Wellborn, Mrs. Edna Wellborn Floyd Booth, a resident of Birmingham, Alabama. Mrs. Booth recalls the site of the mill, and remembers that the Carlton Wellborn homeplace was nearby, as is his now unmarked grave.

Robins Air Force Base, the economic focus of life in northern Houston County in the most recent half-century, now stands near the site of Wellborns Mills, the focus of life in North Houston in its first half-century. The name of Carlton Wellborn is written proudly on the early documents of the county; Wellborn Road is a reminder of the distinguished family who led the area in those pioneering days.

---

[1]Sandy Run Creek forms the present southern boundary of Robins Air Force Base. The southeast corner of Robins is Land Lot 80; Land Lot 81 lies directly east of Land Lot 80; thus the mill was on that portion of land between the Base and the Ocmulgee River along Sandy Run Creek.

Robins Air Force Base is located in the northern end of Georgia's Coastal Plain. It will be seen, as this narrative progresses, that geography would play a major role in choosing the site of the base that would become Robins.

Cemetery containing the graves of Henry Feagin and his grandson, Jim Bob Tharp.

The Feagin house, also known as the "Newberry" house, after its later owners, is seen in the right center foreground. The house was sold and moved from Robins AFB in 1966.

Lee S. Trimble. He was one of the four men most instrumental in the new depot being located in Middle Georgia. The energetic Trimble was manager of Macon's Chamber of Commerce.

Henry and Missouri Feagin: Original Settlers

In the local meaning of the term, an "old-timer" at Robins is a person who reported for duty during the 1940s. (There were less than 200 old-timers still on civilian duty, according to a special listing provided the History Team in February 1982.) It's tough to think of anything here that's older than the Second World War.

But a serene, beautiful spot between the Child Care Center and the NCO Open Mess quickly dispels that notion, for it contains a cemetery dating from a time earlier even than an earlier war: *"The War"*, according to Southerners, whose land was despoiled in the Civil War ending in 1865. Marked by a marble slab, a vault-type grave there reads, "Sacred to the Memory of Henry Feagin. Born Sept 8th 1795. Died April 2nd 1842."

A second *ante bellum* marker nearby states, "Here lies little Jim Bob son of Wm. & Matt Tharp. Died Dec. 17, 1858." Who was Henry Feagin, and who was little Jim Bob?

Henry Feagin was the original settler of the land on which the Robins Air Force Base Community Center and nearby structures now stand. He purchased this tract, a total of 202 1/2 acres, for $600 in 1832.

He and his wife Missouri had five children, four sons and a daughter, who lived in the house he constructed on this property before his death at age 46. He was a planter of means, possessing 1500 acres and 19 slaves, according to the records. His oldest son, Judge George M. T. Feagin, was well-known in Houston County and a long term occupant of the family home.

His great-granddaughter, Mrs. Kay Holland, is employed on base in the Directorate of Distribution. Mrs. Holland has shared family papers which tell of the stirring events of 1865, when the Yankees had captured the President of the Confederacy, Jefferson Davis, near the village of Irwinville in Irwin County south of Houston. On the trip north, as they followed the road from Hawkinsville to Macon, the Yankee guard regiment and the captured Confederate chief and his family spent the night near the Feagin home. Two of the Davis daughters accepted Judge Feagin's offer of hospitality of lodging for the night. The Feagins sent refreshments to the former President and his wife. The events of that night were recalled by President Davis many years later as he was making a tour of Georgia.

The property remained in the Feagin family until the early 1930s, when it was sold to the Newberry family, from whom the property was acquired subsequently for inclusion in Robins Air Force Base.

Little Jim Bob was the son of William and Martha Feagin Tharp, who was laid to rest by the grave of his mother's father upon his death in 1858. Today the graves of children of military personnel assigned to Robins Air Force Base join these two in sharing the peaceful place near the site of the original chapel on Robins Air Force Base.

Robins Air Force Base—The Origin

Despite its geographical assets, Robins Air Force Base would never have been built in Middle Georgia without two essential ingredients—the determination of Macon civic leaders and the assistance of Congressman Carl Vinson. In 1936, under the leadership of Mr. Lee S. Trimble, the Macon Chamber of Commerce embarked on a program of acquiring new industries for Macon. The city had recorded little growth in the previous twenty years. At first only small projects were attempted but as the war clouds gathered and the nation rearmed, Macon saw its chance. It adopted the policy of offering free land for defense installations. Three early successes included a naval ordnance plant (operated by Reynolds Metals), an infantry replacement training center (Camp Wheeler), and an Army Air Corps pilot training facility (Cochran Field).

J. Ellsworth Hall, Jr., assistant city attorney for Macon. He was present when Congressman Vinson told a general in the Corps of Engineers that the depot site selection would be made on merit alone. Mr. Hall's father was a personal friend of Mr. Vinson.

While on a visit to Washington in January 1941, Mr. J. Ellsworth Hall, Jr., Assistant City Attorney for Macon, and Mr. Ruffin G. Chestney of the Georgia Power Company visited Congressman Carl Vinson. Mr. Vinson represented the Sixth Congressional District which included Bibb[2] but not Houston County. Mr. Vinson told them of the War Department's plans to build a new Air Corps Depot somewhere in the southeastern United States. At their request, Mr. Vinson telephoned a Colonel Kennedy of the Air Corps to ask that a survey party be sent to Macon. On February 21, 1941, a delegation led by Congressman Robert C. Word Ramspeck of Atlanta, called on the Assistant Secretary of War for Air, Colonel Robert A. Lovett, to urge that the depot be located at Ellenwood, twelve miles southeast of Atlanta. Mr. Vinson wired Mr. Trimble on February 25, 1941, to advise that he had asked for a "board of inspectors" to be sent to Macon. *The Macon Telegraph* reported on March 7, 1941, that "Atlanta promoters are very confident that city has the inside track."

The survey board arrived in Macon on March 13, 1941, after beginning its tour in Atlanta. Other Georgia sites on the itinerary were Albany, Dublin, Milledgeville, and an area between Cordele and Vienna which had been suggested by Georgia's senior senator, Walter F. George, a native of Vienna. Out-of-state sites included the Nashville and Chattanooga areas in Tennessee. The survey board was composed of representatives from the Army's Air Corps, Ordnance Department, Quartermaster Corps, and Corps of Engineers. A local delegation led by Mayor Charles L. Bowden and composed of representatives from Macon, Bibb County, and the Macon Chamber of Commerce, conducted the tour. The potential sites were Herbert Smart Airport, a farm on U.S. Highway 41, the Bateman peach farm opposite Cochran Field, and the farmland opposite the unincorporated village of Wellston in northern Houston County. The Wellston site, with its railroad and expanse of flat open land was the only one with which the Survey Board didn't find fault. The Maconites were convinced, rightly or wrongly, that the Air Corps wanted the Atlanta site with its big city conveniences and that the air depot had already been promised to Atlanta. It must have raised eyebrows when the Maconites boldly stated that the depot could be built here six months quicker and a half-million dollars cheaper than in Atlanta. The assertion was apparently of interest to the engineers.

Mayor Charles L. Bowden. He led the delegation which accompanied the Army survey board on its tour of potential depot sites in the local area.

The next news received by Macon authorities was that the choice was between Ellenwood and Wellston and that Atlanta businessmen were taking options on the Ellenwood property. A government surveying party was said to have visited Ellenwood in April. During May, Macon secured options on the Wellston site although *The Macon News* reported on May 19, 1941, that the Wellston site was second choice. Mr. C. B. "Boss" Watson of Wellston obtained most of the required options, suggesting prices authorized by Mr. Calder Clay, Chairman of the Bibb County Commis-

[2]The City of Macon was and is the County Seat of Bibb County.

The "Wellston site" as it appeared on September 20, 1941. View is southeast across the airfield area.

sion. On May 23, Mr. Clay and Mayor Bowden sent a letter to General Arnold stating that 1,800 acres of land were already under option at Wellston; that a maximum of 2,500 acres would be donated (the Air Corps had planned for an installation of half that size—1,250 acres); and that local authorities would assume the cost of any condemnation proceedings.

Congressman Carl Vinson. He had represented the U. S. Congressional District, which included Macon, since 1914. His telegram to The Macon Telegraph *stated, "I have done my part."*

In the meantime, the Army requested more information on the Wellston site. A group consisting of Colonel Robert Elliott, Major Francis M. Zeigler (who would become the first deputy commander of the new depot; the Zeigler Housing Project was named for him), Captain Gerald Ward, and Mr. Thomas Lear, a civil servant, arrived on May 24. Except for Major Zeigler, who represented the Air Corps, all of them belonged to the Corps of Engineers. They went over the Wellston area and discussed technical information with the Macon and Bibb County authorities and with Mr. D. William Brosnan of the Southern Railway System.[3] The railroad had promised to equalize freight rates to offset Atlanta's advantage.

On June 8, 1941, Colonel Lovett wrote Congressman Vinson that the Wellston site would be selected if Macon could provide assurance, by the fourteenth, that two thousand housing units would be built for the influx of workers. Mr. Vinson wired *The Macon Telegraph*, "It is now up to the 'Sleeping Giant'. . . . If it cannot meet this requirement, then the depot will be located at Ellenwood. . . . I have done my part. Now it is up to Macon."

A Macon delegation was in Washington on June 13, 1941, with financing commitments and Federal Housing Authority guarantees for not two thousand but three thousand housing units. The Wellston site was approved the next day. Mr. Vinson sent a telegram on June 16, outlining details of the project.

Macon had finally won but obviously part of the story is missing. Something happened between the end of April and the eighth of June to make Wellston the preferred site. It is now known from documents declassified in 1960 that the Site Selection Board recommended the air depot be built at Ellenwood and recommended Wellston as second choice. This could not have been a unanimous decision. Otherwise, it would have received routine approval instead of ending up in Colonel Lovett's lap as a problem to be resolved. The dissenting voice surely came from the

[3]The railroad through Wellston was called the "Georgia, Southern and Florida," but had been a part of Southern Railway since 1895.

Corps of Engineers. The Robins Air Force Base record states that "officials of both cities were aware of a rift between the Air Corps and the Corps of Engineers." These two branches of the Army were organizational equals until March 1942. With both Atlanta and Macon congressmen looking over his shoulder, Colonel Lovett had to have a selection he could defend and he found it. The rock upon which Ellenwood fell was cost and topography. The land there had a market value of $80 per acre and required grading. The land at Wellston was free and flat. According to Colonel Lovett, the Wellston site was selected because the depot could be completed there six months sooner and $1,500,000 cheaper than at any other place considered. In retrospect, the decision made was the correct one. It was a bargain for the U.S. taxpayer and speed of completion would assume an unforeseen importance after December 7, 1941. With the growth of air traffic congestion around Atlanta, the distance from that metropolis would ultimately prove to be a blessing.

The land donated by Macon for "one dollar and other good and valuable consideration" was in two separate deeds and totaled 3,108.5 acres. The land was paid for by increasing business license fees and ad valorem taxes in Bibb County. The Federal Government never came by to pick up the deeds although they were offered. The deeds were finally presented in a special ceremony on May 28, 1943. By then, at least $20,000,000 had been spent on construction and the depot had been operational for eight months—on land for which the Federal Government had no formal proof of ownership.

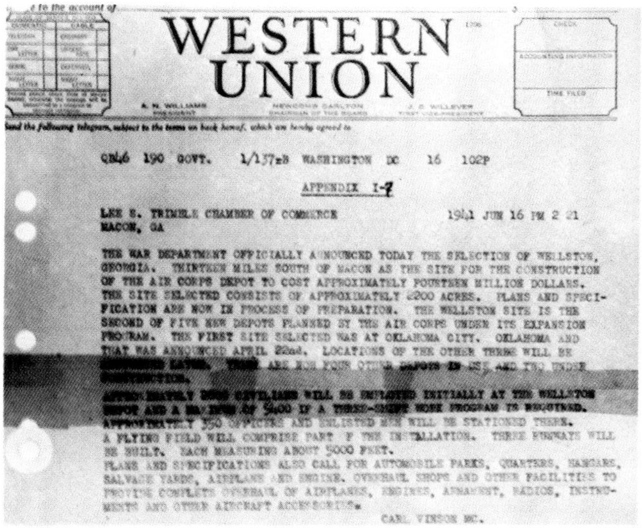

WESTERN UNION

Send the following telegram, subject to the terms on back hereof, which are hereby agreed to

QB46 190 GOVT. 1/137=B WASHINGTON DC 16 102P

APPENDIX I-7

LEE S. TRIMBLE CHAMBER OF COMMERCE 1941 JUN 16 PM 2 21
MACON, GA

THE WAR DEPARTMENT OFFICIALLY ANNOUNCED TODAY THE SELECTION OF WELLSTON, GEORGIA. THIRTEEN MILES SOUTH OF MACON AS THE SITE FOR THE CONSTRUCTION OF THE AIR CORPS DEPOT TO COST APPROXIMATELY FOURTEEN MILLION DOLLARS. THE SITE SELECTED CONSISTS OF APPROXIMATELY 2200 ACRES. PLANS AND SPECIFICATION ARE NOW IN PROCESS OF PREPARATION. THE WELLSTON SITE IS THE SECOND OF FIVE NEW DEPOTS PLANNED BY THE AIR CORPS UNDER ITS EXPANSION PROGRAM. THE FIRST SITE SELECTED WAS AT OKLAHOMA CITY. OKLAHOMA AND THAT WAS ANNOUNCED APRIL 22nd. LOCATIONS OF THE OTHER THREE WILL BE ANNOUNCED LATER. THERE ARE NOW FOUR OTHER DEPOTS IN USE AND TWO UNDER CONSTRUCTION.

APPROXIMATELY [illegible] CIVILIANS WILL BE EMPLOYED INITIALLY AT THE WELLSTON DEPOT AND A MAXIMUM OF 5400 IF A THREE-SHIFT WORK PROGRAM IS REQUIRED. APPROXIMATELY 350 OFFICERS AND ENLISTED MEN WILL BE STATIONED THERE. A FLYING FIELD WILL COMPRISE PART OF THE INSTALLATION. THREE RUNWAYS WILL BE BUILT. EACH MEASURING ABOUT 5000 FEET. PLANS AND SPECIFICATIONS ALSO CALL FOR AUTOMOBILE PARKS, QUARTERS, HANGARS, SALVAGE YARDS, AIRPLANE AND ENGINE. OVERHAUL SHOPS AND OTHER FACILITIES TO PROVIDE COMPLETE OVERHAUL OF AIRPLANES, ENGINES, ARMAMENT, RADIOS, INSTRUMENTS AND OTHER AIRCRAFT ACCESSORIES.

CARL VINSON MC.

Telegram announcing the selection of Wellston as the site of an Air Corps depot. The original telegram was the property of the Macon Chamber of Commerce and no longer exists. This is a copy made at Robins Field during World War II.

The "Good Old Days"

Mrs. Helen Marshall Tucker

Mrs. Helen Marshall Tucker, currently an employee of the 2853d Civil Engineering Squadron, was born on what is now Robins Air Force Base and lived there until she was fourteen years old. She is one of the four daughters of the late Philip Marshall. The family home was on the site of what is now Building 220. For the purpose of this history, she was asked to describe the "Wellston site" before the construction crews came in late 1941. The following paragraphs are based on that interview.

Directly across the dirt road, now Second Street, in front of Philip Marshall's house was a wheat field. Behind the house was a pasture where his dairy cattle grazed. The family sold its milk to Dixie Dairies in Macon. The cows drank from a pond in the vicinity of where the big hangar, Building 125, now stands. Beyond the pasture, adjoining it on the north, were fields of watermelon, cantaloupe, and peanuts. The home of Mrs. Tucker's grandfather, Bob Marshall, was further east on what is now Second Street in the vicinity of Buildings 309 and 310. That explains the pecan trees at the north end of those buildings. The flat land north of this point, including the site of the Air Freight Terminal, was a cotton field.

Perhaps nothing so graphically illustrates how tiny the community of Wellston was than the fact that Bonaire was bigger. The Wellston children went to school at Bonaire. Wellston had but one store while Bonaire had two. You could ride the seven miles to Bonaire on the railroad for a nickel. If you didn't have five cents, you could ride anyway if you promised to pay it back the next time.

So much has been written about the dismal poverty of the South before World War II that later generations tend to think that the misery was universal. That, of course, is a false picture. The Marshalls, for example, were certainly not wealthy but neither were they uncomfortable. Mrs. Tucker remembers her childhood as happy and carefree.

Carl Vinson
1883—1981

Carl Vinson was born on a farm near Milledgeville, Georgia in 1883—one of eight children. He graduated from Mercer University Law School in Macon in 1902. When he went to Congress in 1914, at age thirty, he was the youngest member of the House of Representatives. Woodrow Wilson was President of the United States, Harry S. Truman was a farmer in Missouri and Dwight D. Eisenhower was a cadet at West Point. When Vinson retired in 1965, he had served in the House longer than anyone else in American history. It was not just time spent. He was one of the few in a true leadership role. President Lyndon B. Johnson described him as a man who "belongs with the Clays, the Websters, the Calhouns . . ." He was the only American to have a U. S. Navy ship (nuclear aircraft carrier) named after him while he was still alive. He disagreed with the statement that he had served under nine presidents—insisting that he had served "with" them.

Carl Vinson was assigned to the House Naval Affairs Committee in 1917. He had never been in the Armed Forces and, before coming to Washington, had never seen a battleship but he did his homework. After becoming Committee Chairman in 1932, he began to press for a naval buildup though the big increases didn't come until 1938-1940. He was considered the father of the two-ocean Navy and was known thereafter at the Pentagon as "The Admiral." His colleagues in the House called him the "Swamp Fox" because he didn't lose many battles. One that he did lose was the plan to fortify Guam. The Japanese took it quickly and easily when the war came. The military liked him. He often took their side against their civilian bosses at the Pentagon but he supported Truman when he fired General of the Army Douglas MacArthur. "There is but one Commander-in-Chief," he reminded the nation.

Vinson became Chairman of the House Armed Services Committee when it was created in 1947. Except for four years when the Republicans controlled the House, he was Chairman of Armed Services until his retirement in 1965. Once when asked to be Secretary of Defense, he replied, "I'd rather go on running the Pentagon from up here."

Carl Vinson was one of the easiest men in Washington to see but he wouldn't always talk. He once told reporters seeking an interview, "I'm too young to play in the snow." And he could be coy. When asked if he would run for a twenty-sixth term, he said, "I never take off my socks until I get to the creek."

Mr. Vinson was much honored in Middle Georgia—"Mr. Robins Air Force Base" being one of his many nicknames. In an unpublished paper on the origin of the Warner Robins depot, written at the request of Robins officials in 1943, Mr. Lee Trimble said of Mr. Vinson, "He is due great credit for his cooperation." Atlanta certainly gave him full credit. Its leaders were bitter—believing that they had been beaten in their attempt to gain the depot by Vinson's enormous political influence rather than by any conceivable logic. That view, of course, carries the insinuation that Mr. Vinson abused his power and it is false. What he did was to keep everybody honest. The Air Corps wanted the Atlanta site because of the attractions of a big city. Once Mr. Vinson became personally interested in the matter, the War Department didn't dare approve the Atlanta site unless it could be justified on economic grounds and it couldn't be done.

The "Wellston site" as construction began. The view is south toward the main industrial area. The building on the left is a temporary repair shop.

Robins Air Force Base—The Construction

The directive authorizing $14,006,150 for construction of the new air depot was received by the Area Engineer, Major Robert H. Elliott, just ten days after the Wellston site had been selected. The cost-plus-fixed-fee contract was let on August 8, 1941, to Griffin, Mion and Shepherd of Atlanta for "Construction of an Air Corps Supply and Repair Depot." Aqua Systems, Inc. of New York City was to install "an aircraft fueling system" and Robert and Company of Atlanta the steam and electric utilities. Since everything was to be built at one time, it was planned as a model layout for a major Air Service Command Depot.

Construction began on September 1, 1941, at the site of the runways. Topsoil removal and some drainage were required. Most of the labor came from the surrounding area. There were 350 workers at the beginning and 1,084 by October 26. After Pearl Harbor there was a frantic effort to speed up defense construction across the United States. The Georgia Air Depot (the unofficial name used by the Corps of Engineers) had 2,200 construction workers by Christmas of 1941. The peak of 6,600 was reached in May 1942. Charles W. Griscom, a civilian with the Corps of Engineers, was in charge of all construction beginning December 1, 1941. He was commissioned as Major in June 1942 and was assigned as Post Utilities Officer as soon as the construction was completed. He stated that there was never a day of labor troubles during the entire project. Common laborers were paid forty cents an hour.

Army Corps of Engineers (Savannah District)

The resident field office of the Savannah District, U. S. Army Corps of Engineers, located at Robins Air Force Base, supervises the construction of major facilities at this installation. The role of the Army Corps of Engineers at Robins can be traced back to the beginnings of the installation in 1941, when Savannah District was designated design and construction agent for all authorized facilities.

Headquartered in Savannah, Georgia, the Savannah District is one of five within the Corps' South Atlantic Division and one of thirty-six corps districts nationwide. The Corps, as planners, designers, and builders, has had its personnel on continuous duty in the Savannah area since 1829 when work was initiated on construction of Fort Pulaski and maintenance and improvements to Savannah Harbor were begun under the Corps' civil works mission. Robert E. Lee, as a Lieutenant, served as acting officer-in-charge of Fort Pulaski's construction, as well as acting Commander of the Corps' Savannah Station—later to become the Savannah District.

Today, the Savannah District is responsible for Army and Air Force design and construction activities within the three-state area of Georgia, South Carolina, and North Carolina. In addition to Robins, it supports some of the largest military installations in the United States—e.g., Fort Bragg, Fort Benning, Fort Jackson, Fort Stewart, and Fort Gordon, as well as Moody, Seymour-Johnson, Pope and Dobbins Air Force Bases. Projects range from multimillion dollar military hospitals, commissaries and barracks to family housing and other support facilities.

Photograph of a metal tag unearthed on Robins AFB in 1982. The inscription on the rim bears the name of the contractor who built the depot—Griffin, Mion and Shepherd. The legend in the center of the tag reads, "Georgia Air Depot, Wellston, Ga."

The site of depot supply buildings (now buildings 300 and 301). The view is northeast, October 22, 1941.

Brigadier General Augustine Warner Robins 1882-1940

The General A. Warner Robins, whose name is memorialized in Robins AFB and the City of Warner Robins, was born in Virginia on September 29, 1882. The facial features which you see here probably differ somewhat from those of the original man. He was in a plane crash in 1921 in which his jaw was broken and "his face so shattered that he was in Walter Reed Hospital for six months." When he died peacefully of a heart attack at Randolph Field, Texas in 1940, Douglas Southhall Freeman wrote an obituary for the *Richmond News Leader*. Dr. Freeman wrote:

"He was a Virginian of the Virginians. His father (William Todd Robins, Colonel CSA) was one of the most daring of all Confederate troopers and was the man put in charge of the advanced guard when 'Jeb' Stuart made his renowned ride around McClellan in June 1862. As for Warner Robins' mother, who doesn't remember affectionately, Mrs. Sally Nelson Robins (author and genealogist), one of the wisest and wittiest of a famous generation of women?"

Warner Robins graduated from West Point in 1907. His first assignment was with the 12th (Horse) Cavalry at Fort Oglethorpe, Georgia. In 1916, he was with the punitive expedition which Gen. John J. "Blackjack" Pershing led into Mexico after Pancho Villa had raided and burned Columbus, New Mexico. In 1917 he left the cavalry for the air service—receiving his wings in August 1918.

After his recovery from the plane crash in 1921, the assignments of Warner Robins were related, for the most part, to what we call "logistics." For four years, beginning in 1931, he was executive officer to the Chief of Materiel Division (Brig. Gen. Henry C. Pratt) at Wright Field, Ohio. For four more years, beginning in 1935, he was chief of Materiel Division with the rank of Brigadier General. He left Ohio in 1939 to become Commandant of the Air Corps Training Center at Randolph Field. It was there on June 16, 1940, that "he finished his military duty, as loyal soldier and servant of his country."

The composite picture of General Robins which emerges from the testimony of those who knew him is that of an able, self-confident, but congenial man who loved to entertain his friends with the tales of his adventures. The favorites concerned his narrow escapes in China and how he managed to avoid arrest. He had gone there in 1910, posing as a millionaire tourist but was actually mapping the country for Army Intelligence. If the general could see the base which bears his name today, he would marvel at how the world has changed.

The first Commander of the Southeast Air Depot (the unofficial name used by the Air Corps) arrived on November 9, 1941. He was a forty-five-year-old Lieutenant Colonel from Atlanta, Georgia, named Charles E. Thomas, Jr. It is a name that looms large in the history of Robins Air Force Base and, though not generally known, in the history of the city of Warner Robins. Colonel Thomas set up his headquarters in a tent at what was then the Macon Municipal Airport—Herbert Smart, which was named for an earlier mayor of Macon. At the dedication of the new air depot in April 1943, Mayor Bowden recalled his first meeting with Colonel Thomas: "I remember when I first saw him—out at the municipal airport. It was raining, and cold, and we talked while sitting on a cot in his tent. Instead of telling us how pleasant it was to be in our midst, he started giving us Hell about the condition of the roads down here and the need of housing. Frankly, he made me mad as a wet hen until I went away and cooled off. I then realized that he was not sent down here to shovel salve on local dignitaries but to do a job."

Three days after Pearl Harbor, Colonel Thomas moved his headquarters into rented office space at 419 Broadway in Macon. Additional space at 415 Broadway was leased on January 6, 1942. From February until May 1942, office space was rented in the Bankers Insurance Building and, for a time, Colonel Thomas had his office there. It was under these makeshift conditions that workers were hired into Civil Service and arrangements made to have them trained at other Air Service Command installations. Administrative employees were trained on the job in Macon. Some 1,300 others were sent to Middletown, Pennsylvania, and Fairfield, Ohio, for training. Also a cadre of 400 specialists were recruited from the depots there. Together, these 1,700 trained employees provided the nucleus of an industrial work force.

It was during this time that the installation taking shape in Houston County was officially named "Robins Field" (January 23, 1942) in honor of Brig. Gen. Augustine Warner Robins. The air depot received its first official name of "Wellston Air Depot" on March 14. On the confusing subject of names, it should be noted that the Base itself has always been "Robins"—first Robins Field and now Robins Air Force Base. The depot has had many official names but they are always prefaced by the geographical place. Since October 1942, the depot name has always been prefaced by "Warner Robins" because the town of Wellston had changed its name to Warner Robins.

It might not have been by chance that the name "Robins" was given to the base under construction at Wellston. In this 1921 photograph taken at Wright Field, Ohio, the officer in the front row, second from left, is Lt. Col. Augustine Warner Robins. The officer in the back row, fourth from left is Lt. Charles E. Thomas, Jr. Thomas served under Robins at Wright Field from 1921—1926. The two men and their wives became close friends and remained so for the rest of their lives. Thomas stated, publicly, in 1943, "I cut my teeth and experienced my growing pains under his fatherly and inspiring influence.

Karl McPherson 1912-1975

Mr. Karl McPherson was born in Cuthbert, Georgia on March 5, 1912 and was graduated from the public school system of that city. He afterward attended Abraham Baldwin Agricultural College, George Washington University Law School, and Woodrow Wilson School of Law.

Mr. McPherson entered civil service in 1935 as an employee of the Federal Power Commission, first in Washington, D. C., and then in Atlanta. He later transferred to the U. S. Civil Service Commission in Atlanta. He came to Macon in 1941 to establish a Board of Civil Service Examiners for the hiring of civilian employees for Camp Wheeler. He could not have known that he would spend the rest of his career in Macon.

In late 1941, Mr. McPherson was assigned the task of hiring civilian employees to staff an installation under construction at Wellston, Georgia. To obtain the thousands of skilled industrial workers required for a depot situated in a primarily agricultural area was a monumental undertaking and the biggest challenge of his career. McPherson, himself, would later admit, "I didn't think we could do it." Four hundred technicians were hired at other air depots and thirteen hundred workers, hired locally, were trained at existing depots before the one at Wellston was completed. Those seventeen hundred skilled workers played a major role in training the others. The Signal Corps also conducted intensive training for its employees at Robins Field and Bibb County established a school where hundreds of engine and aircraft mechanics were trained. As McPherson would later recall, "We tore apart and reassembled one old single-engine aircraft so many times that it hung loose when it was assembled."

Mr. McPherson, though he hired the first Robins employees and would hire more than 100,000 in his career, was not an employee of Robins Field in the beginning. He left the Civil Service Commission to become Civilian Personnel Officer at Robins Field on February 16, 1942. He was Chief of Civilian Personnel until his retirement on February 27, 1970 as well as Deputy Director of Personnel from the time the position was created in 1945. Mr. McPherson died on November 17, 1975.

The story is widely told that many of the early employees of Robins Field thought, that since McPherson hired them, he must be the big boss. When asked where they worked, they would say, "at Mr. McPherson's airfield." And, in a sense it was.

Supply activities of the new air depot began first in a steel warehouse at Herbert Smart Airport and later in a building on Waterville Road in Macon. A rented warehouse at 306 Bay Street in Macon was used to house Quartermaster supplies.

The first building to be completed on Robins Air Force Base no longer exists. It was a wood frame complex facing Fourth Street. The headquarters of the Security Police Squadron now stands on the site. The Area Engineer moved his office there from Cochran Field on October 10, 1941, although the building was still under construction and not ready for occupancy until January 1942. This precedent would be repeated.

The bulk of construction was north of Second Street. It included three five thousand feet runways, flight operations, the maintenance hangars, and such structures as engine repair, engine test and Post Headquarters—now the location of the Executive Offices of the Directorates of Maintenance and of Contracting and Manufacturing (Bldg 220). It was completed and occupied in May 1942. The major construction south of Second Street was the two Depot Supply Warehouses—now Buildings 301 and 300 except for the parts with two stories and a basement which project out in front of the original structures. These were added in 1954. What is now Building 301 was pressed into service on April 14 while still under construction. A tarpaulin stretched over steel girders served as a roof for the office staff. With supplies pouring in and no forklifts, it was not uncommon to see officers in shirt-sleeves helping to unload railcars.

Foundation of Aircraft Repair, Building 125, northwest corner, January 21, 1942.

This complex of four connected buildings in the second block of Fourth Street was the first structure to be completed (January 1942). It was a complete plant with administrative offices, barracks, cafeteria, and dispensary. The Area Engineer occupied the unit on the right on October 10, 1941.

Installation of concrete storm sewers, December 30, 1941.

Armament Fire Control Building. It now constitutes the south end of Building 162. The view is northwest, January 28, 1942.

Paint, Oil, and Dope Storage, Building 327. The view is northwest, February 27, 1942.

Radio Repair Building on left, Equipment Repair Building on right. The site is now occupied by Building 158. The view is northwest, February 12, 1942.

Post Headquarters, Building 220, staked-out. The view is northwest, February 12, 1942.

West wing of Depot Supply, Building 301. The view is south, March 9, 1942.

Chemical Storage, Building 306. The view is north, March 9, 1942.

Paving the northeast-southwest runway. The view is northeast, March 9, 1942.

Excavation for steam plant, Building 177. The steel framework on the left is Engine Repair, Building 169. The view is south, April 3, 1942.

Second Street. The view is west on April 3, 1942.

"A" Street, showing Quartermaster Buildings on the right. The view is north, April 10, 1942.

Post Headquarters, Building 220. The view is northeast, April 10, 1942.

Aircraft Repair, Building 125. The view is west, March 20, 1942.

Engine Storage Building 173. The view is northeast, April 3, 1942.

Signal Corps Warehouse, Building 165. The view is northeast, April 24, 1942.

Engine Test, Building 181. The view is northeast, April 24, 1942.

Aircraft Repair, Building 125. The view is northeast, April 24, 1942.

Field Officer Quarters, April 24, 1942.

Non-Commissioned Officers Quarters, April 28, 1942.

With construction of the depot underway, the Air Corps decided to add an additional function—troop training. Directly south of and adjacent to the Macon-donated land, it purchased 2,637 acres and leased 782 more for a military cantonment and training area. Construction of the cantonment area began in May 1942. It was built in the pattern of a dispersed airdrome as one would expect to see in a combat zone.

Landscaping and camouflaging of Robins Field began in late summer of 1942. Mr. William L. Young of Macon was, in those days, a captain in Post Utilities and was involved in the landscaping. He didn't understand why time and money should be spent on it. Colonel Thomas told him, "When the war is over, they are going to be closing bases. I want this one to be too pretty to close." Camouflage was never taken very seriously and was stopped in August 1943. One experiment—painting the runway dark green—was too successful. It was necessary to sprinkle yellow sand over it so the pilots could see where to land.

It is difficult to assign a specific date when construction ended and operations began because they overlapped. However, the cantonment area and all of the depot necessary to perform the mission had been completed by August 31, 1942. The permanent recording of Maintenance production began thirty-one days later (October 1, 1942).

The contention that an air depot could be built quickly at the Wellston site was proven by the event, for it established a record. The original contract called for completion in thirty months. It was done in twelve. This was largely due to the impetus of Pearl Harbor but is still remarkable. Of the four Air Service Command Depots under construction during this time, Wellston was started last and was finished second.

Depot Supply, Building 301. The view is south, May 18, 1942. Note concrete foundations for water tower in foreground.

Engine Repair, Building 169. The view is north, May 18, 1942.

Control Tower and Operations Hangars, Building 110. The view is southeast, May 25, 1942.

Civil Engineering

The Civil Engineering Organization was activated as "Construction and Utilities Office" during construction of Robins Air Force Base in 1942. Although the name and organization structure have changed many times, civil engineering functions and responsibilities have remained the same—construction, maintenance, and repair of physical facilities and operation of utility services.

Operation of the base's first utility systems and maintenance of the new facilities were vested in the Post Utilities Officer. He was assigned to the installation commander's staff; however, technical direction was exercised by the District Engineer of the Corps of Engineers. As early as 1945, the function was staffed with about 500 personnel of various crafts to perform the maintenance, repair, and utilities work. On July 16, 1945, the base fire department was transferred from the Provost Marshal to the Construction and Utilities Office.

The Construction and Utilities Office was changed to Air Installations Office on August 7, 1945. In 1946, the Army Air Forces, for the first time, assumed technical supervision over maintenance activities. With approval of the National Security Act of 1947, the Department of the Air Force was established. This act transferred maintenance, real estate, and construction functions from the Army to the Air Force.

The Air Installations Office assumed responsibility from the Federal Public Housing Authority (FPHA) for maintenance of military family housing on January 1, 1948. There were 1,058 units at the time.

The role of the Civil Engineering mission changed dramatically in 1965. The organization was realigned to perform direct combat support to the Air Force mission by developing and maintaining a highly skilled, mobile combat military engineering force capable of rapid response for worldwide contingency operations. This new military force was named Prime BEEF (Base Engineer Emergency Forces). It was originally staffed with 144 military positions, but has now been expanded to 286.

Aircraft Repair, Building 125. The view is northeast, June 4, 1942.

The Great Patriotic War

The conflict which began with the invasion of Poland in September 1939 and ended with the surrender of Japan in September 1945 is known in America as "World War II." In the USSR it is called the "Great Patriotic War." It was a time when, to everyone's surprise, the Soviet peoples laid aside their terrible internal dissensions and joined together in a common effort to save "Mother Russia." But in the United States, it was also a great patriotic war. On December 7, 1941, the "date of infamy," United States forces and territory were victims of a surprise and unprovoked attack by the Japanese. That made the war popular—the most popular in all of American history. There was a tremendous outpouring of patriotism. It was a time that brought out the best in Americans. People thought more in terms of "us" and "ours" and less in terms of "me" and "mine." For that reason, the era is still remembered with a great deal of nostalgia. To be sure, we had our slackers, shirkers, and opportunists even then but they were few and they were scorned. Those first few months brought heartbreaking reverses but the public felt anger, not despair. No one ever doubted the eventual outcome. The United States of America had just begun to marshal its mighty resources.

When the permanent recording of maintenance production began at the Wellston Air Depot on October 1, 1942, the U.S. Marines were nailing down their conquest of Guadalcanal and the epic Battle of Stalingrad was underway. In a little more than six weeks, an American Army would go ashore in North Africa. The tide had begun to turn.

For one who has known Robins Air Force Base only in the decades which came after the 1940s, the contrast with Robins Field of World War II is startling. Robins Field was an Army installation. In addition to the Army Air Forces (name adopted in the War Department reorganization of March 9, 1942), many other segments of the U. S. Army were represented. Chemical Warfare Service, Ordnance Department, Quartermaster Corps, and Signal Corps all conducted supply operations separate from the depot's Supply Division. All were involved in troop training. Signal Corps, by far the largest of these four organizations, had an extensive maintenance operation where it

The Fifth Street Gate in 1942. (The main gate, Second Street, is on the dust jacket.)

Robins Officers' Wives Club

Originally called "Robins Field Women's Club," the Robins Officers' Wives Club was organized by Mrs. Charles E. Thomas, Jr., wife of the first commander. The first meeting was held on November 17, 1942. The purpose of the club is to "promote friendship among its members, to further their social and intellectual well-being, and to coordinate their combined efforts in welfare, educational, cultural, and social activities." In connection with its stated purposes, the Robins Officers' Wives Club sponsors numerous activities for its members. Since it began, the club has either sponsored or contributed to a series of welfare, community service, and social improvement projects too numerous to list. Among its first projects was establishment and operation of a nursery for the children of mothers employed at Robins Field during World War II. *Robins Chirps*, the official publication of the club, began in 1944 as a stenciled newsletter. Since September 1963, it has been a printed magazine. Current welfare projects are funded by profits from the "Thrift Shop," which opened in 1952 and is operated by club volunteers.

Officers' wives Tea in late 1940s

Post Headquarters, now Building 220, wearing its camouflage, 1943.

overhauled air and ground radar as well as communications and navigation equipment. The Signal Corps was particularly proud of having designed and built several items of equipment for use in the field, e.g., marker beacons and portable test benches. (Perhaps this activity should be regarded as the forerunner of the Warner Robins Depot as the "Avionics Center of the Air Force.") The Chemical Warfare, Ordnance, Quartermaster, and Signal organizations were reduced to Staff Offices in November 1944 when their operational functions and personnel were transferred to three divisions of the depot (Maintenance, Supply, and Personnel Training).

When the depot at Robins Field became fully operational on October 1, 1942, it was part of the Air Service Command which had its headquarters at Patterson Field, Ohio. The Air Service Command was the largest command in the Army Air Forces. The old World War II slogan, "Keep'em flying," describes its mission as well as any three words in the English language. The Headquarters exercised control through four Air Service Area Commands. Their geographical areas of responsibility coincided with the geographical areas of the four Continental Air Forces. The Air Service Area Command Headquarters for the Southeastern United States was in Atlanta. The depots at Robins Field, Georgia, and Brookley Field, Alabama, were in the same geographical area. Each depot was assigned specific sub-depots over which it exercised "command and technical control." A sub-depot was an Army installation with flying activity. Most of this flying activity was flight training because the United States was in the process of building the greatest air arm that the world had ever seen. Air Service Command personnel at the sub-depots did

This photograph, taken after the Signal Corps supply function was transferred to the depot in 1944, shows the size of the operation. The building is now the Physical Sciences Laboratory, Building 165, at the corner of B and Second Streets.

Four Signal Corps Officers at Robins Field, 1943. Left to right, *Maj. Roy. J. Kelly, Capt. John A. Trask, Capt. Jack H. Braucht, Capt. Emil Patureau. Braucht was a Second Lieutenant in October 1942 when he received orders to report to Robins Field, Wellston, Georgia. He asked his roommate where Wellston was. The roommate, who was from Macon, said he had never heard of it. After the war, Braucht continued his career at Robins Field as a civilian. He retired in 1975 as a Deputy Division Chief in the Directorate of Materiel Management. He also continued his military career, retiring from the Air Force Reserve as a Colonel in 1979.*

Robins Air Force Base Exchange

The Robins Air Force Base Exchange traces its roots to Herbert Smart Airport but its beginning at Robins Field was in May 1942 in the Post Headquarters building (Bldg 220). Lt. Verne N. Hamilton, Post Exchange Officer, hired Mr. Pierce Davenport to assist in establishing the Exchange at Robins Field. Mr. Davenport advanced twenty dollars out of his own pocket to start the branch in the Post Headquarters building. Its assets consisted of a soda fountain, cigarettes and candy. The operator was Mrs. Billie Yaughn.

The Main Exchange was opened in Building 250 on June 19, 1942. There were thirty-two branches by April 1944, consisting of retail, food, service station, supermarket, bake shop, bowling alley, skating rink, beauty shop and barber shops. The Main Exchange was located in Building 250 from 1942 until August 11, 1971, when the new Main Exchange Shopping Complex opened in Building 914. Branches of the original Exchange have consolidated, organized and reorganized over the years. The Exchange operated under Base Services from July 1960 until the inception of integrated management in January 1972. It currently operates under the West Georgia Area Exchange at Fort Benning, Georgia.

Post Exchange, 1943.

what we now call "field maintenance" and requisitioned their supplies from the depot to which they were assigned. The depot issued supplies to its sub-depots and did the more complex work which we now call "depot maintenance."

On February 1, 1943, the four Air Service Area Commands were abolished and the United States was divided into Control Area Commands administered by the various depots. Thus the Warner Robins Depot received its own geographical area consisting of South Carolina, Georgia, and Florida east of the Appalachicola River. It would be pointless to recount the sub-depots which were assigned to the Warner Robins Depot at one time or another. Not only was the number constantly changing, but on January 1, 1944, the Air Service Command began the transfer of most of the sub-depots and their personnel to the numbered Air Forces and the commands which they were servicing. A better measure of the workload of the Warner Robins Depot was that its geographical area contained approximately 6,500 aircraft which it supported from the standpoint of supply and depot maintenance.

It should be understood that the air depots of World War II were general depots restricted to a geographical area. Every geographical area had aircraft which used sparkplugs so every depot stocked sparkplugs. The same concept was true for maintenance. The Warner Robins depot maintained whatever aircraft were assigned in its geographical area. Some of the famous aircraft on which Warner Robins depot employees worked were the B-17, B-24, B-25, B-26, and P-51.

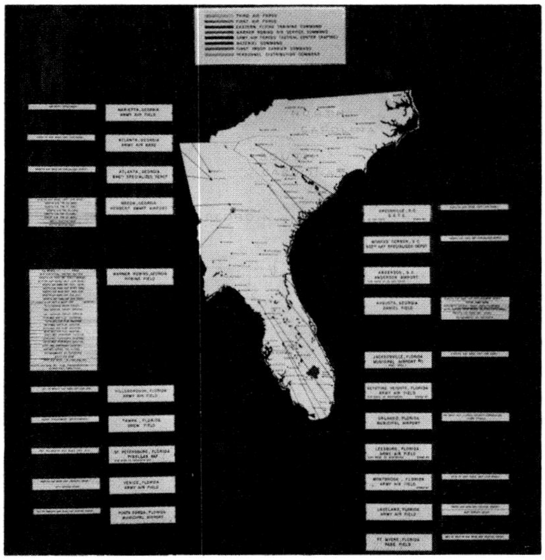

The geographical area of responsibility of the Warner Robins depot after the addition of North Carolina in August 1944. By this time, jurisdiction over most of the sub-depots had been transferred to the numbered air forces and to other commands.

Cochran Field, Macon, was an early sub-depot of the air depot at Robins Field.

Grace Martin, early depot employee, in her blue, civilian uniform. The uniform was not mandatory and its use soon died out. Interviews with early depot employees indicate that the uniform was more popular with female than with male employees.

Christmas luncheon, 1942, Identification and Investigation Unit. Note women in foreground wearing civilian uniform.

Machinist apprentice graduating class, 1943. Classes were held at Lanier High School in Macon. The high school also provided the instructors.

Identification and Investigation Function, 1942.

The first big project of the depot's maintenance function was the major modification of fifty-seven B-25s. The report which reviewed the first year of maintenance operations didn't say what the modification was but that it was completed ahead of schedule and "in a most satisfactory manner." The second major project was the modification of P-51 and A-36 aircraft. Neither the number of aircraft nor the nature of the modification was mentioned, but it was completed "on schedule." The report also noted that a new record was established with the overhaul of eight hundred aircraft engines in the month of August 1943. The report contained one curious paragraph which should not go unnoticed:

> On August 19, 1943, the Maintenance Division set up a specific ten minute rest period to be observed twice daily by all civilian personnel. This procedure is believed to have eliminated much lost time previously used in smoking, visiting Coca-Cola machines, etc.

Another major mission of the depots was the training of the military personnel who would "keep 'em flying" in the combat zones. Actually, there was more involved in the troop training at Robins Field than the maintenance and supply of the Army Air Forces. In addition to Air Depot Groups and Air Service Groups, numerous medical platoons as well as military police, quartermaster, ordnance, chemical, and signal companies were trained. Records on the exact number of troops trained at Robins Field during World War II were not maintained and the reported totals included Air Service Groups trained at the sub-depots through 1943. Also, four Air Depot Groups were trained at Herbert Smart Airport in Macon before the cantonment area at Robins Field was occupied in September 1942. With these facts in mind, a total of 50,000 troops trained at Robins Field during World War II appears to be a reasonable estimate. The Fourth Air Depot Group left Herbert Smart Airport for Australia less than two weeks after Pearl Harbor. The Fifth Air Depot Group left for England in March 1942. The 36th and 37th Air Depot Groups went to North Africa. The 37th took part in the action at Casablanca, Morocco. The 38th Air Depot Group was the first unit to complete its training on Robins Field. It left Robins Field in December 1942—going first to Oran, Algeria, and later to Naples, Italy. The inevitable confusion which existed in 1942 as the United States rushed headlong to turn the tide of war has long been forgotten. A glimpse of it can be seen in the letter which Lt. Col. Russell Scott, Executive Officer in Charge of Troops, at Robins Field wrote to the Adjutant General on September 30, 1942:

> This Headquarters . . . receives a large number of men who are assigned to units that have not been activated at this station, to units not at this station, or to units that have left this station.

The gigantic U. S. Army Air Forces of World War II was not created overnight. The installations in the United States played the major role in the development and maintenance of the infrastructure necessary to man, launch and sustain the vast air armadas which hammered central Europe incessantly from middle 1943 onward.

Aircraft repair, 1943.

Civilian training, Robins Field Classroom, 1943.

Aurie Akin operating a drill press, 1942.

In the severe labor shortage of World War II, women were trained and successfully performed jobs which had always been thought to be masculine. In the aircraft engine repair function, for example, the civilian workforce was more than sixty percent female in 1945.

End of the dayshift, 1943.

Troop training at Robins Field, 1944. Air Service Group under simulated strafing attack.

Enlisted men's barracks, 1943. Different colored shingles were used to achieve camouflage effect.

Chemical Warfare training, Robins Field, 1944. Soldier without mask is Mervin DeCastro.

Security Police

The first of many military police companies was activated at Wellston Air Depot on September 15, 1942. Shortly thereafter, a Civilian Guard Force was established. During World War II the Guard Force provided post security and also patrolled the highway (now Georgia Highway 247) north and south for approximately 20 miles in each direction. In addition, the Guard Force policed the town of Warner Robins using an old barracks style building on Second Street as the jail.

In 1963, the Civilian Guard Force was abolished. Assigned personnel were required to pass a written test and physical examination to qualify for the newly established police positions. Those not passing the written test or physical examination were reassigned to other positions on the installation outside of the police field.

The first of many civilian positions in the 2853d Security Police Division were converted to military in 1979, under the Critical Military Skill Program. By 1980, significant changes occurred in the unit's organizational structure and mission.

On March 12, 1980, the 2853d Security Police was redesignated as a squadron. Then, in October the squadron was tasked to establish the capability to deploy personnel to perform internal base defense and security in support of the USAF War Mobility Plan.

In 1981, the 2853d Security Police Squadron was selected for the fourth consecutive year as the outstanding unit in the Air Force Logistics Command. It was selected first runner-up in 1980, as the outstanding small Security Police Unit in the Air Force.

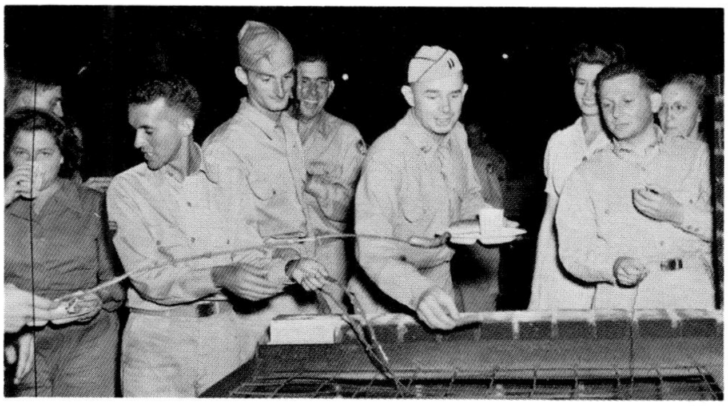

Wiener roast following religious service, post picnic ground, October 1944.

Troop Review, Robins Field, 1943.

Depot Commander and Staff, May 1944, Left to right, First row: *Col. Russell Scott, Col. Shelton E. Prudhomme, Brig. Gen. Charles E. Thomas, Col. Hansford W. Pennington, Col. Victor H. King, and Maj. Charles D. Jacobs.* Second row: *Maj. Conrad C. Schatte, Lt. Col. Morris Martin, Maj. Clyde W. O'Hern, Lt. Col. Joseph E. Cheek, Maj. George W. Brown, Maj. Asa M. Dodd, Maj. Leslie C. High, Maj. Maxwell S. Strong, and Maj. George F. Warren.* Third row: *Lt. Col. William F. Niedernhofer, Lt. Harriet N. MacLaurin, Col. Thomas E. Wood, Maj. George A. Miller, Maj. Earl V. Wolfe, Lt. Col. Henry C. Urquhart, Maj. William L. Daves, Lt. Col. Don R. Keefe, Lt. Col. Charles M. Dittrich, Lt. Col. Ralph B. Shaw, and Maj. Maurice J. Barron.* Fourth row: *Maj. Daniel J. Duval, Maj. Pierce F. Davies, Col. Paul J. Mathis, Maj. David St. P. Dubose, Lt. Col. Frank T. Hanby, Maj. George W. Scott, Maj. Warren C. Bailey, and Capt. Roy J. Kelley.*

Parade through downtown Macon, April 26, 1943. The parade followed ceremonies at Robins Field, in which the Warner Robins Army Air Depot was formally dedicated. After the parade, further dedication ceremonies were held at the Macon Municipal Auditorium.

B-17s drop their deadly cargo.

The Housing Problem— Birth of a City

For the Wellston Air Depot, housing was perhaps the most difficult problem of World War II. It had seemed so simple in the beginning. Macon had been asked to build two thousand housing units to accommodate the expected population increase which the air depot would bring. In its exuberance it had promised three thousand, though it didn't believe so many would be needed. The Federal Housing Authority had estimated the requirement at only 1,500. These houses would not be reserved specifically for depot workers; they would be a part of the total housing available in Macon and Bibb County.

Colonel Thomas met with the Macon Chamber of Commerce shortly after his arrival in Middle Georgia in early November 1941. He listened to the report of progress and he wasn't impressed. What he said must have left the Maconites stunned. Not only was the available housing insufficient but it was being built in the wrong place. He wanted it across the highway from the depot—not in Bibb County but in Houston. In fact, he wanted a planned city for civilian workers with movie theaters, drugstores, fire protection, and sidewalks.

Nineteen days after Pearl Harbor, Mr. Lee Trimble of the Macon Chamber of Commerce wrote Colonel Thomas a letter on the housing situation. He noted that construction of the air depot had been accelerated but he was optimistic. He cited statistics from the U. S. Labor Department that Bibb County had a net gain of 1,242 private dwelling units in 1941 plus 719 Federal Project units. Also, Macon builders would continue their efforts so long as the demand existed. The builders were so deeply committed that he feared they would be badly burned if the building site was shifted to Wellston. Further, he pointed out, Wellston was not incorporated and Houston County had no zoning authority. If a planned city was necessary, it would be best to build it opposite Cochran Field, in southern Bibb County, where it could serve both installations. Colonel Thomas, whose opinion was based on personal observation, was not persuaded by the statistics. If he had been, the depot would have been crippled at the outset for lack of housing. It nearly was. The flaw in the statistics was that general economic activity was increasing and more jobs were becoming available. Also, the other defense installations in the area were expanding beyond their original plans and the extent of this increase was classified information. Hence, the population of Bibb County was increasing rapidly. What looked like an ambitious building program was scarcely keeping the county's head above water.

Col. Charles E. Thomas, Jr. at his desk, March 1943. The first depot commander (November 9, 1941-May 21, 1944), Thomas was a native Georgian. The City of Warner Robins was born out of his struggle to secure local housing for depot personnel.

With a freeze on the sale of automobile tires shortly after Pearl Harbor, their rationing in January 1942, and the projected strength of the depot rising, Colonel Thomas was a worried man. He could no longer delay seeking a Federal Housing Project for Wellston, but he continued the search for private capital to build his city. The H. J. McKeon Company of Miami was interested, and so was the Fickling and Walker Company of Macon. Elaborate plans for a community were submitted to the War Production Board, but fell victim to wartime scarcity. Colonel Thomas would have to settle for a subdivision around which a city could grow. In the meantime, Mr. Ben H. Butts, Jr., of Macon bought the McKeon Company and turned it into a holding company which bought two hundred acres of land, the site of the subdivision called "Robins Manor." The construction contract was let to the Wellston Housing Company which had been incorporated by Fickling, Walker, and Butts. Pipe, copper wire, and bathroom fixtures were now in critical short supply. To make a long story short, the National Housing Authority was finally prevailed upon to approve construction of the first 250 houses in Robins Manor. Notification came to Colonel Thomas in a confidential teletype dated June 16, 1942. It outlined the "devious means" by which priorities for building materials might be obtained. A transcript was furnished to the Wellston Housing Company enabling it to begin construction on July 30. The village of Wellston announced five days later that it would change its name to Warner Robins.[4] The name became official on September 1, 1942. The depot became the Warner Robins Army Air Depot on October 14, 1942, and has, since that date, always been prefaced by the name "Warner Robins." On October 17, 1942, approval was granted to build another 250 homes in Robins Manor. The little town of Warner Robins was off and running. It was incorporated on March 5, 1943, and Mr. C. B. Watson took the formal oath of office as Mayor on March 15, 1943.

Robins Manor, 1943 - Construction of this first subdivision began on July 30, 1942. Five days later, local residents announced their decision to adopt the name "Warner Robins".

The name change from Wellston to Warner Robins became official on September 1, 1942.

The effort to secure a Federal Housing Project for Wellston was not progressing entirely satisfactorily to Colonel Thomas. He wanted family housing but the War Department had a policy that only civilian barracks would be built for defense workers. In a letter to Colonel P. A. Feringa of the Corps of Engineers, dated May 29, 1942, Colonel Thomas argued his case. He stated that the projected civilian strength of the depot was now in excess of 13,400 and the single men were going off to war. Married persons could not be forced to leave their families and live in barracks. "After all," he wrote, "we are still living in a democracy."

The civilian barracks or dormitories reserved for white workers were built across the highway from Robins Field on 139.9 acres of land bought with War Department funds. These buildings were due west of the big maintenance hangar. Government policy after Pearl Harbor required them to be at least one thousand yards from a military target. Two buildings reserved for black workers were built across the highway from the Ninth Street Gate on land already owned by the Federal Government as part of the cantonment area purchase. Most of the dormitories were built in the shape of a cross but a few had only three wings. The standard arrangement was 43 beds to each wing—a total of about 2,500 beds. There were two private rooms with private baths in each wing. The other forty-one persons shared two bathrooms. Construction began in August 1942 and ended in January 1943 but the buildings were occupied as they were completed. The dormitories served a crucial need but were not completely successful. The number of vacant beds never dropped much below one thousand.

The continuing effort to secure federal family housing for Warner Robins did succeed with Lanham Act funds being obtained to construct 65 buildings with 498 family units. (The Lanham Act was named for Rep. Fritz G. Lanham, D-Tex, who sponsored the legislation in 1940.) The project was an instant success with the first units occupied in June 1943 and construction completed in December 1943. Another 560

[4]The Robins Field version of the name change was that the townspeople did it themselves in order to identify with the base—thus exonerating Colonel Thomas from any involvement. He had argued against the depot being named after the insignificant hamlet of Wellston. When he was overruled by Air Service Command Headquarters and it became the Wellston Air Depot anyway, his only alternative was to persuade the townspeople to change the name of the town. The Watson brothers were quoted in the August 16, 1942 issue of *The Macon Telegraph* as follows:

> But we are willing to do anything to help our country. We can offer no objections to anything Colonel Thomas wants us to do. While we hate to see the name of Wellston lost, we feel that it is right to thus honor a man who meant much to the air forces of the United States.

units were built in early 1944. The entire project, at Colonel Thomas' suggestion, was named "Zeigler Place" in honor of Colonel Francis M. Zeigler who died from injuries sustained in a plane crash a mile and a half south of the base on December 2, 1942. He had been assigned to Robins Field in September 1942 as Executive Officer to Colonel Thomas. Colonel Zeigler, in his official capacity at Air Service Command Headquarters, had been involved in the planning of the depot from the beginning, and is said to have been the one who designed its physical layout.

A typical civilian dormitory, late 1942.

The Watson Brothers, 1942 - Left to Right, Clint K., Charles B. "Boss", and Walter C. Clint was justice of the peace, in charge of gas rationing, and a member of the local draft board. "Boss" was postmaster and later mayor. Walter was a retired farmer and former county tax receiver.

North Zeigler Place under construction, 1943. The view is northeast. North Zeigler was on the north side of Watson Boulevard between Myrtle Street and Commercial Circle.

Private construction also continued with the "Warner Robins Defense Homes" (one hundred houses and two hundred duplex units) being built by the Ellinor Company of Atlanta, adjoining the Commercial Circle quadrant west of Davis Drive and south of Watson Boulevard. The first units were occupied in November 1943 and the project was completed by the end of April 1944. Fickling and Walker, which managed Robins Manor, offered to manage the Defense Homes. The suggestion was submitted to Brigadier General Thomas for comment. (He had attained the rank in September 1943. His permanent rank was still Lieutenant Colonel.) He objected on the grounds that "competition is needed at Warner Robins" and the idea was dropped.

Wellston or Warner Robins was not immune to the vice which has traditionally followed boomtowns and military encampments. It arrived with the troops and civilian workers and set up shop across the highway from Robins Field. Beer and wine were the only legal alcoholic beverages in the county. Long before the town was incorporated, local residents, led by "Boss" Watson, formed a committee to curb the rowdiest of these establishments. Their weapon was "economic blockade." An Improvement Association was also formed by some of the property owners to prevent the sale of land to such businesses. This effectively limited the sale of alcoholic beverages to First Street. The Improvement Association was assisted by the Legal Department at Robins Field. A report prepared by the Office of Community War Services in March 1944 stated that "The town is remarkably free of 'honky-tonks' and does not give the impression of having an approximate population of 10,000." Before the coming of Robins Field, it had had a population of forty-seven.

South First Street, City of Warner Robins, 1943. The name of the theater was not changed until 1954.

Brigadier General Thomas left Robins Field in May 1944 and went on to become a Major General. He returned in 1950, to what was now Robins Air Force Base, as Commander of the Fourteenth Air Force. To his list of credits one more can be added: "Founding father of Georgia's eighth largest city."

At the dedication of the Warner Robins Army Air Depot on April 26, 1943, Mayor Bowden mentioned early misunderstandings between Colonel Thomas and the Macon leadership. He didn't say what they were. The Robins Air Force Base records leave no doubt that the misunderstanding was over housing. When Colonel Thomas first met with the Macon Chamber of Commerce and proposed a planned city in Houston County, he felt he was doing the local real estate interests a favor. He could have ignored them and sought a mammoth Federal Housing Project at Wellston. When they showed no interest in his plan, it almost looked like ingratitude. The problem with the Macon real estate interests was that they weren't convinced the demand was there. The fact is, they did respond and with great urgency when they finally saw that the need was not only real, but critical. Colonel Thomas had his own reasons for preferring housing built with private capital. It is obvious from his letters as well as his public and private remarks that he had a dream which went beyond World War II. His base would be permanent, and the city was a part of that dream. Public housing could be abandoned and demolished when the war ended. It would not be so easy to abandon extensive private housing.

These early misunderstandings were resolved and left no ill feelings. Mayor Bowden went on to say in his dedicatory address:

> *Time has now proved him right . . . I am not throwing unjustified bouquets when I say that our country . . . is fortunate in having this man, Colonel Charles E. Thomas, Jr., in its service. In passing, let me say that I . . . find that he is a nice fellow so long as you do what he says in advancing the work he was sent here to do.*

Overhead view of Commercial Circle, looking southwest, in 1949. The houses in the upper right of the photograph were part of the Warner Robins Defense Homes, sometimes called "Ellinor Homes." The building in the lower left was the commercial building constructed as part of the Zeigler housing project. The house amid the trees was the home of "Boss" Watson.

Charles E. Thomas III School - This first school in Warner Robins was built with a Federal grant of $100,000 and was completed in 1944. The annual salary of teachers ranged from $850 to $1,250. The school was named for the elder son of Colonel Thomas. The son was a West Point graduate and 2nd Lieutenant in the Army Air Forces. He was "lost at sea" on a training flight in April 1942.

Postscript: The Last Goodbye!

Major General Charles E. Thomas, Jr. retired as Commander of the Fourteenth Air Force at Robins AFB on January 31, 1955. The City of Warner Robins said its last goodbye ten days earlier. The ceremony began with a parade of two thousand troops and musicians up Watson Boulevard. *The Macon Telegraph* reported the event as follows:

A retiring general stood in a wintry wind yesterday as the city he helped to build . . . gave him his proudest salute.

It was a thunderous, yet warm and tender tribute from Warner Robins to Maj. Gen. Charles E. Thomas, Jr.

General Thomas cited the young city's "phenomenal start," and warned, "we must not lose that momentum."

"We must," he added, "continue to improve by building better schools and streets; maintaining a progressive city government . . ."

He called for an immediate program . . . "for attracting a type of industry that will complement our military program."

General Thomas said he had found his "reward in these memories, in the great honor you have accorded me today and in the knowledge that we have built a good city . . ."

It was a memory-stirring moment when two "old-timers"—General Thomas and aging C. B. "Boss" Watson shook hands after the ceremony and posed for pictures.

Said Watson to the old friend who named him Warner Robins' first mayor: "I don't see how you held the tears back." The Air Force general dropped an arm around the shoulder of his fellow pioneer and answered, "It's hard to leave . . ."

This 1945 aerial photograph (the view is east) shows the various housing projects discussed in this section and their geographical relationship. The subdivision in the lower right, south of Watson Boulevard, is the Warner Robins Defense Homes. The one across Watson Boulevard, to the left of the Defense Homes, is Robins Manor. The cross-shaped buildings above Robins Manor are civilian dormitories. The long buildings above both Robins Manor and the Defense Homes are Zeigler Apartments. The old Wellston Community was clustered in the vicinity of the railroad depot. The new town was built on farmland to the west. This photograph shows a gap of almost two city blocks between the two.

Victory and Uncertainty

At the end of March 1945, Allied armies were across the Rhine and the war in Europe moved inexorably toward its close. The manpower complement of the Warner Robins depot was 23,670—10,686 military and 12,984 civilians. This was more than four times the number of civilians and more than thirty times the number of military which had been expected to staff the depot in the beginning. When V-E Day came on May 8, there were brief ceremonies in the various work areas. The end was not yet but big changes were already underway. By June 30 the military strength of the depot was down to 6,578—a drop of more than 4,000 in three months.

Japan accepted the allied terms of surrender on August 14, 1945, and a ceasefire went into effect but the formal surrender was not signed until September 2, 1945 (V-J Day). Upon cessation of hostilities on August 14, Maintenance immediately placed 1,500 civilian workers on two weeks leave. When that period expired without a formal treaty, one thousand of them were put on thirty day furlough. In the meantime, on September 1, the work week for Robins employees was reduced from forty-eight to forty hours. After V-J Day, the official manpower ceilings came down from higher headquarters requiring an immediate reduction in force. By the end of September 1945, the strength of the depot was 15,733—a decline of 4,430 military and 3,507 civilians since the heady days of March. This first sharp and sudden reduction of civilian jobs was not as traumatic as it might appear. The manpower reduction which would follow the end of the war was expected. Large numbers of employees took other jobs and resigned—not waiting for a reduction in force. In those organizations in which V-J Day didn't bring an immediate drop in activity, there were complaints that the workers were disappearing faster than the workload. But this was only the beginning of the decline of Robins Field and its air depot. By the end of September 1946, the strength of the depot was 163 military and 5,104 civilians—a decline of more than eighteen thousand personnel in eighteen months and the all-time low was yet to come.

Robert Byrd, employee in Propeller Shop, March 1945. At that time the Allied armies were across the Rhine. Byrd later served as a Warner Robins City Councilman and as Chairman of the Houston County Commission. He was serving on the Commission during the time of the land exchange of the early 1970s, a continuing evidence of the support rendered by the local governments to secure the future of Robins Air Force Base.

Germany surrendered, unconditionally, on May 7, 1945.

Corrosion Control, May 1945.

Officers in the Supply Division, June 1945.

After the second atomic bomb was dropped (Nagasaki, August 9, 1945), Japan offered to surrender on condition that the emperor be permitted to retain his throne. The allies agreed and a ceasefire went into effect on August 14, 1945.

Engine Repair, May 1945.

The formal Japanese surrender was signed on the Battleship Missouri in Tokyo Bay, September 2, 1945. General of the Army Douglas MacArthur signed for the victorious allies.

Army Air Forces Day, August 1946. Unidentified spectators gaze into the bomb-bay of a B-25.

The post World War II employment low for the depot was reached by the end of March 1947. It was 202 military and 3,959 civilians. But, of course, nobody knew that this would be the low point. Numerous houses stood vacant as the little town of Warner Robins withered. The peanut patch which is now the site of the Miller Hills subdivision could have been bought for less than $50 an acre. No doubt the thoughts of many turned to "what if." For example, the deeds for the land which Macon had given to the government for "one dollar and other good and valuable considerations" had additional clauses. The deeds also provided that title would revert to the City of Macon in fee simple if ever "the War Department, by and through the then qualified and acting Secretary of War, shall declare the aforesaid real estate to be surplus to the needs of the military authorities, and shall for a period of one year discontinue the use of said real estate for military purposes." Not only the real estate but all "appurtenances and improvements thereon" would revert to the City of Macon. The land which Macon had donated was the site of the airfield and the main industrial area—the most valuable part of the installation. It would have been a princely return on Macon's investment of $97,011.67 which is an average of $31.20 an acre.

Earlier accounts of the history of Robins Air Force Base give scant notice to this era. It was a time of neither glamor nor optimism. Still, several items from the period are worthy of note. On September 18, 1947, the U. S. Air Force became an autonomous branch of the U. S. Armed Forces—formed from the personnel, equipment, and facilities of the Army Air Forces. Five months later (February 16, 1948), Robins Field became Robins Air Force Base, the name which it bears today. Even before that (July 2, 1946) the name of the depot was changed from Warner Robins Air Technical Service Command (WRATSC) to Warner Robins Air Material Area (WRAMA). This was the seventh name of the depot but the one which it would

Advertising the Maintenance picnic and dance, 1947. Woman on left, standing, is Juanita Bryant.

Depot Commander and Staff, August 1946. First row, left to right, *Col. B. J. Peck, Col. F. P. McCue, Col. G. W. Polk, Col. R. V. Ignico, Col. B. B. Cassiday, Col. E. W. Fox, Lt. Col. L. C. Gatto.* Back row: *Col. E. T. Noyes, Col. W. O. Moore, Lt. Col. G. A. Bosch, Maj. R. A. Huckaby, and Capt. H. R. Daniels.*

have for the next twenty-eight years. Though the depot name was changed to Warner Robins Air Logistics Center on April 1, 1974, the term "WRAMA" still slips out occasionally in conversation.

There were other changes from this era which would endure far into the future. For example, in January 1946, the Locality Wage Rate Plan for all ungraded or wage board employees was implemented—a system still in use today. Prior to that time, workers in the industrial skills were paid on a national wage scale instead of according to the prevailing wage rates in the local area. Of the 3,879 Robins Field ungraded civilian workers at the time, almost fifty-one percent were over the scale and there was much misgiving about this plan. However, as implemented, the greatest loss suffered by any employee did not exceed sixteen dollars per year.

Some memorable depot maintenance projects of the era involved storage of some of the massive quantities of war material left over from World War II. Thousands of vehicles were prepared for storage and the base looked like a giant parking lot. Hundreds of B-29s were filled with bags of silica gel and sprayed with a strippable plastic popularly called a "cocoon." These were removed from storage and recommissioned during the Korean War.

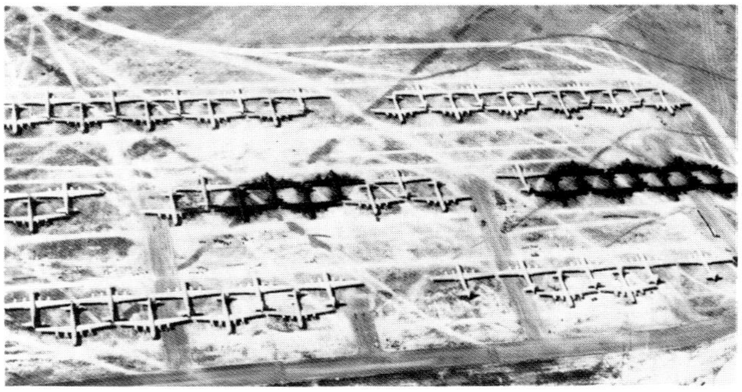

B-29 Cocoon Project, 1947. The plastic was sprayed on the aircraft in several applications. Each successive layer was a different color so the operator could tell when he had applied each coating uniformly.

Second Street Gate at shift change. Photo is dated May 14, 1947. Note the absence of Building 210 and 215. They were built in the 1950s.

Engine Test, June 1948. Eugene Kinser, Joe B. Ballard, George Felver, and George Davenport (left to right) *have attached the wooden propeller to test-run the engine. Kinser was foreman of the Engine Repair Shop in which 1800 people were employed.*

Supply conference, October 1947. The visitors have completed a tour of the cocooned B-29.

As stated in a previous section, the air depot of World War II was a general or self-sufficient depot confined to a specific geographical area. The modern depot concept, which has existed since the early 1950s, is that of a specialized depot with Air Force-wide responsibilities. The late 1940s was a time of experimentation which provided the bridge of transition to the new depot. The first step in this direction came in late 1947. The Warner Robins depot was selected, for no apparent reason, to aggregate materiel for the establishment of a "readiness reserve" for the Strategic Air Command (SAC). SAC operated worldwide and thus, in this case, the depot broke the bonds of its geographical area. The success of this program led, in 1948, to the service testing, followed by the adoption, of the "two-zone" concept of logistics support. Warner Robins was one of the two depots selected for the service test. Under the two-zone system, the United States was divided at the Mississippi River. Air Force stocks of a given stock number were then divided between a western depot and an eastern depot. Each depot, for its assigned stock numbers, supplied all Air Force installations in its zone plus the nearest overseas theaters. For example, the western depots supported Air Force activity in the Pacific and the eastern depots did the same for Europe. This system had been implemented for many, though certainly not all, stock classes at the time of the Berlin Airlift (June 1948 to September 1949). One of the two-zone stock classes, which had been assigned to the Warner Robins depot at the time, included sparkplugs. The Warner Robins depot repaired approximately 150,000 sparkplugs per month—of which 50,000 per month were furnished to sustain the aircraft involved in the Berlin Airlift. The experience was one element of proof that the idea of a specialized depot was feasible.

In February 1947, a C-45 crashed into the Ocmulgee River swamp upon take-off from Robins Field. Seven persons, all from what is now Wright-Patterson Air Force Base, Ohio, were killed. Because of the impenetrability of the swamp, it took fourteen hours to reach the crash site, though it was only a mile and a half from the runway. This led to the so-called "Swamp Trails" project. Twelve miles of trails were cut into the Ocmulgee swamp, on the east side of the base, so that crash sites could be reached more quickly.

The long twilight for Robins Air Force Base would not end in the 1940s. That was destined to come in the early 1950s with the second build-up of the depot, the Warner Robins Air Material Area, to the World War II level and, in many respects, to even greater heights. However, the late 1940s did provide the first glimmer of the new dawn. In October 1949, the Headquarters, Fourteenth Air Force moved to Robins from Orlando Air Force Base, Florida. Except for the depot itself, Robins Air Force Base had its first major tenant[5] organization, the story of which is found in the next chapter.

Aggregation of Readiness Reserve materiel for SAC, 1948. The depot, in this instance, was not restricted to its assigned geographical area.

C-45 crash in Ocmulgee River Swamp, February 1947.

[5]The American Red Cross has been active at Robins Air Force Base since the beginning and still is, performing numerous educational and humanitarian functions. District 6 Headquarters of the Air Force Office of Special Investigations (AFOSI) was located on Robins Air Force Base in January 1948. In the reorganization of September 30, 1972, it was redesignated as AFOSI Detachment 0712, the name which it still bears in 1982.

Crash rescue team, February 1947. It took 14 hours to reach the crash site.

Alonza Hassell "Hal" Cotton 1921-

Mr. A. H. Cotton was born at LaFayette, Alabama on July 23, 1921. He graduated from Clay County High School, Ashland, Alabama. He attended the University of Alabama and George Washington University.

Mr. Cotton began his Civil Service career on February 8, 1941 in the supply function of the 97th Sub-Depot at Craig Field, Selma, Alabama. From September 1943 until April 1944, he was assigned to the Oklahoma City depot. From April 1944 until May 1947, Mr. Cotton was Deputy Director of the Supply System at the Atlantic Overseas Air Materiel District, Newark, New Jersey. At Newark, he worked on the procedural control of wartime requisitions to overseas theaters, the inactivation of in-transit depots and later, on the transfer of property to the War Assets Administration.

Mr. Cotton came to Warner Robins in May 1947 as the Deputy Director of Supply and Services. He served in that capacity until the creation of what is now called the Directorate of Materiel Management on January 1, 1959. From that date, until his retirement on January 4, 1980, he was Deputy Director of Materiel Management.

Mr. Cotton's career of thirty-nine years spanned the history of what we now call the Air Force Logistics Command from its formative years through the evolution of the command into what it is today. From the time of his assignment to the Warner Robins depot in 1947 until his retirement in 1980, Mr. Cotton played a towering role in that evolution. As an early example, he was directly associated with the establishment of the Readiness Reserve of the Strategic Air Command—the program in which the depot first broke the bonds of its geographical area. Three times, the Air Force awarded him the Decoration for Exceptional Civilian Service, the first time at the age of 31. In 1967, he became the first employee of the Air Force Logistics Command to receive the Department of Defense Distinguished Civilian Service Award. On May 13, 1981, he was presented the President's Award for Distinguished Federal Civilian Service, the highest honor that can be bestowed upon a civilian employee of the federal government. Mr. Cotton remains, today, one of the nation's greatest living authorities on the logistics of the United States Air Force.

Swamp Trails Project, 1948.

The "Swamp Weasel," 1948. It was one of several vehicles tested for crash rescue in the Ocmulgee swamp.

Work-saving invention, April 1949. John L. Durden, Civil Engineering, demonstrated his device which enabled one person to lift, remove, and install a Chrysler air conditioning compressor.

Maintenance apprentice trainees, July 1949.

A luncheon in the Directorate of Supply, 1949.

Robins Field—As It Was In April 1945

Looking west over the Ocmulgee River swamp, Robins Field, and the City of Warner Robins. The western limit of the city was Pleasant Hill Road.

Looking northeast over Warner Robins Defense Homes and Robins Manor toward the airfield. The street in the foreground is Pleasant Hill Road.

Looking northeast over headquarters building and industrial area.

Looking northeast over what is now Building 300.

Looking northeast over Operations Building (110).

Looking northwest over what is now Buildings 309 and 310.

Looking northwest over what is now Building 300.

Looking south over Officers Club toward cantonment area.

Looking east over the cantonment area. Photograph was apparently taken from near the site where Building 640 now stands. The major intersection on the right is now Ninth Street and Robins Parkway.

Looking northeast over Theatre No. 2, Service Club and cantonment area. The intersection in the center of the picture is now Seventh Street and Robins Parkway.

Looking west over South portion of cantonment area.

Looking southwest over Quartermaster laundry—now used as a Plastics Shop.

A PHOTOGALLERY: *Leaders of the 1940s*

Commanders, Air Depot at Robins Air Force Base

Brig. Gen. Charles E. Thomas, Jr.
November 1941—May 1944

Charles E. Thomas, Jr., was born at Atlanta, Georgia, April 8, 1896. He left Georgia Tech after three years to become a flying cadet in October 1917. He graduated from the School of Military Aeronautics at the University of Texas in April 1918. He was commissioned in the Air Reserve in August 1918 and in the Air Service of the Regular Army in July 1920.

Thomas became depot adjutant of the Fairfield, Ohio, Air Intermediate Depot in March 1922. In May 1927, he became assistant supply officer for the air depot at Camp Nichols, Philippine Islands. He became supply officer at the San Antonio Air Depot in May 1929 and operations officer at Maxwell Field, Alabama, in July 1933.

In March 1941, Thomas was assigned to the Provisional Air Corps Maintenance Command at Patterson Field, Ohio. He assumed command of the Southeast Air Depot at Wellston, Georgia, in November 1941. In May 1944 he became deputy chief of staff of the 21st Bomber Command in the South Pacific and in October of that year was appointed deputy commander for operations of the AAF in the Pacific. He became Chief, Aviation Advisory Group to the Chinese Government in June 1947, and in January 1949 was appointed Commanding General of the 1503d Air Transport Wing at Tokyo, Japan. In August 1949 Thomas became Inspector General of the Continental Air Command at Mitchel Air Force Base, New York. It was there that he was promoted to Major General. On July 17, 1950, he was appointed Commander, Fourteenth Air Force, Robins Air Force Base, Georgia, in which position he retired from the Air Force on January 31, 1955.

General Thomas died at San Antonio, Texas on March 17, 1969. His widow, Gladys Whittington Thomas, died at San Antonio in April 1982. One son, Charles E. Thomas, III, died at sea during World War II. The surviving son, Robert William Thomas, lives in San Antonio and is employed at the San Antonio Air Logistics Center.

Maj. Gen. Thomas H. Chapman
August 1944—August 1945
June 1952—November 1952

Brig. Gen. Robert V. Ignico
December 1945—June 1952

Thomas H. Chapman was born in Michigan. He attended high School in Detroit, and went to the University of Michigan. He attended the Air Corps Tactical School in 1934.

An enlisted man before he entered commissioned ranks, General Chapman was a sergeant in the enlisted reserve at Camp Wheeler, in 1917 to 1918, going to Officer Candidate School at Camp Johnston, Florida in 1918. He was commissioned August 8, 1918, and served as Motor Transportation Officer at Camp Humphries, Virginia and the Presidio of San Francisco from 1918 to 1920.

Transferring to the Air Corps in 1921, General Chapman received his flying training at Carlstrom Field, Florida, and Kelly Field, Texas, receiving his wings at Kelly in 1922. He remained at Kelly as an instructor until 1925 when he moved to Duncan Field, as engineering officer of the San Antonio Air Depot, a post he held until 1928, and to which he returned in 1930 after a two-year tour of duty in the Philippine Islands.

In 1935 he went to Wright Field, Ohio, to serve as Chief of Procurement Inspection of the Materiel Command. In 1942 he became Deputy Commander of the Second Air Force at Spokane, Washington. In 1943 he was transferred to Salina, Kansas, as Wing Commander.

From Salina he moved to San Antonio, Texas, as Deputy Commander of the San Antonio Air Depot, a post which he held until coming to Robins. General Chapman retired from his position as WRAMA Commander in November 1952. General Chapman is deceased.

Robert V. Ignico was born on October 14, 1895 in Boston, Massachusetts. He received his LL.B. degree from Washington and Lee University in Lexington, Virginia in 1917. He graduated from the Advanced Air Service School, Kelly Field, Texas in 1923.

Ignico was one of the officers who, early in the 1920s helped organize Air Corps Logistics. The effort was headed by Major A. Warner Robins, later Brigadier General Robins.

In 1924 Ignico served as commanding officer, Philippine Air Depot, and in 1929 he became commanding officer, 10th Airship Company, Langley Field, Virginia. From June 1932 to October 1936 he was Depot Supply Officer at the San Antonio Air Depot. In 1937 he was assigned as Supply and Maintenance Officer, Mitchel Field, New York.

In November 1941 he was assigned to the Air Service Command at Wright Field, Ohio. From July 1942 to October 1944 he was Commanding General, Eleventh Air Force Service Command, operating in the Aleutians, and Deputy Commander of Eleventh Air Force. In September 1944 he became Commanding General, Spokane Air Service Command, Spokane, Washington.

General Ignico was a rated balloon pilot, combat observer, balloon observer and aerial observer. He is married to the former Arlene Davidson and they have two children. He is retired and lives in Belleair Bluffs, Florida.

Deputy Commanders, Air Depot at Robins AFB

Col. George W. Polk, Jr.
December 1945—May 1948

Col. Edward W. Fox
June 1948—August 1948

Col. William C. Sams
August 1948—November 1951

Biographical Notes on the Deputy Commanders

Colonel Francis M. Zeigler, September 1942—December 1942, (not pictured): Colonel Zeigler was the son of Mrs. Lina Spann Zeigler of Denmark, SC. He was born in Allendale, SC on July 18, 1902. He was graduated from Clemson College and was a senior pilot with 2900 flying hours to his credit. He was first commissioned in October 1928 through the AUS. Prior to entering the service, he was a manufacturing engineer for General Electric. He was one of the planners of this depot, having much to do with its organization while he was stationed at the Fairfield, OH depot. He was transferred here in September 1942. He died of injuries sustained in a plane crash which occurred two miles south of the Depot in December 1942. He was survived by his wife, Mrs Milfred Van Ausdel Zeigler, and two children. He is buried in Arlington National Cemetery.

Colonel Russell Scott, December 1942—July 1944 (not pictured): Col. Scott had been director of military and other troop training since moving to Warner Robins from Herbert Smart where he was the commanding officer of the 4th Station Complement. Col Scott was educated in Cleveland, OH public schools and Emerson Institute, majoring in business administration. On August 18, 1942, he was assigned to Wellston Air Depot as "Executive Officer in Charge of Troops."

Colonel James W. Brown, Jr., September 1944—March 1945 (not pictured): Colonel Brown was the former commanding officer of Spokane Army Air Field before reporting to Robins Field in August 1944. He is a native of LA and received his military training at the U.S. Military Academy, West Point, graduating in the class of 1928. He received his basic flying training in 1929 at Brooks Field, TX. He went later to Kelly Field for advanced training. In 1931, he enrolled in the Maintenance engineering course at Chanute Field and now holds a Senior Pilot's rating. Col. Brown had served a tour of duty from 1934-1936 as a squadron operations officer, Albrook Field, Panama.

Colonel Polk: Colonel Polk was reassigned as Comptroller of Robins until his departure in March 1948.

Colonel Sams: Colonel Sams was reassigned as Commanding Officer, 75th Air Depot Wing, Kelly AFB, TX on 6 December 1951.

Directors/Deputies, Air Depot at Robins AFB

Unfortunately, photographs for some leaders were not available. These leaders are, however, mentioned in the "Biographical Notes" section.

Lt. George E. Kegin
Director of Supply Division
November 1941—April 1943

Col. Elmer H. Jose, Jr.
Director of Supply Division
April 1943—February 1944

Col. Robert F. Cullwell
Director of Supply Division
February 1944—July 1944

Lt. Col. W. H. Funston
Director of Supply Division
July 1944—September 1945

Lt. Col. Mason N. Grower
Director, Deputy of Supply
September 1945—February 1946

Col. Ben J. Peck
Director of Supply Directorate
February 1946—August 1949

Col. Len B. Dietrickson
Director of Supply and Services
August 1949—August 1951

Mr. Alonza H. Cotton
Deputy Director, Supply and Services
May 1947—November 1958

Maj. Charles A. Day
Assistant Engineering Officer in Charge
August 1942—September 1942

Col. S. E. Prudhomme
Depot Engineering Officer
September 1942—June 1944

Col. H. W. Pennington
Maintenance Division Chief
June 1944—July 1944

Col. G. W. Noland
Maintenance Division Chief
July 1944—February 1945

Col. James H. Reed
Chief of Maintenance
February 1945—October 1945

Col. Henry A. Sebastian
Chief of Maintenance
October 1945—June 1946

Col. William O. Moore
Chief of Maintenance
June 1946—May 1947

Col. Maurice M. Beech
Chief of Maintenance
May 1947—February 1948

Col. Howard A. Moody
Director of Maintenance
February 1948—July 1951

Mr. E. B. Griffin
Deputy Director, Procurement and Production
1945—1953

Col. George W. Polk, Jr.
Chief, Personnel and Training Division
1946—May 1948

Col. Edward W. Fox
Chief, Personnel and Training Division
August 1948—January 1949

Lt. Col. George A. Bosch
Chief, Personnel and Administration
May 1948—August 1948

Mr. Karl McPherson
Chief of Civilian Personnel and
Deputy Director of Personnel
February 1942—February 1970

Col. George W. Polk, Jr.
Comptroller
January 1948—March 1948

Lt. Col. Ralph M. Fawcett
Comptroller
March 1948—December 1950

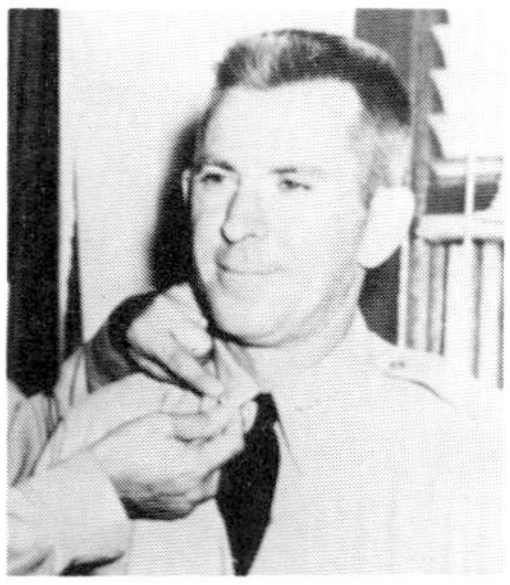

Lt. Col. Sidney B. McNair
Deputy Comptroller
1947—February 1952

Biographical Notes on the Air Depot Directors/Deputies

Lt. George E. Kegin: The first director, Lt. Kegin set up the supply function at the Atlantic Ice and Coal Company warehouse on Waterville Road in Macon. He moved to Depot Supply No. 1 building (now Building 301) on April 14, 1942. Rising to the rank of Lieutenant Colonel, he retired in Anchorage, AK as an insurance representative.

Col. Elmer H. Jose, Jr.: The Air Freight Terminal was constructed during Colonel Jose's administration, an invaluable facility for the distribution of supplies and equipment during World War II.

Col. Ben J. Peck: Colonel Peck remained in the logistics discipline, serving as Director of Supply and Services at Sacramento AMA, Deputy Chief of Staff/Supply at Hq. USAF, as Assistant to General Hefley at Chateauroux in France, Director of Supply at San Antonio AMA, and as Director of Materiel Management at Mobile AMA, a position from which he retired. He lives in Tampa, FL.

Col. Len B. Dietrickson: Following his tour at Robins, Colonel Dietrickson was Commander at Memphis Air Force Station. While on a visit with Robins friends at Lake Sinclair, he died unexpectedly.

Mr. Alonza Hassell Cotton: Mr. Cotton became the Deputy Director of the new Materiel Management function in 1958, a position from which he retired to his home in Perry. He serves as consultant on a part-time basis, continuing his interests in gardening native azaleas and in church and civic affairs.

Maj. Charles A. Day: During his tour, the first airplane, an A-20, was worked for minor repair.

Col. S. E. Prudhomme: Under Colonel Prudhomme's leadership, Maintenance expanded greatly in support of World War II. Major workloads included maintenance/modification of aircraft, repairing/testing engines, and overhauling/repacking parachutes.

Col. H. W. Pennington: Colonel Pennington is retired and lives in San Antonio, Texas.

Col. G. W. Noland: At one point during this period, fourteen hundred planes were here for storage of which over three hundred were for strategic reserve.

Col. James H. Reed: Maintenance shared vehicle workload with Supply during his tour. Colonel Reed is retired and lives in Nashville, TN.

Col. Henry A. Sebastian: During this time, Robins assumed long-time storage of some 350 B-29 aircraft. Aircraft repair and vehicle maintenance workloads increased under his leadership.

Col. William O. Moore: Under his direction, 192 A-26 aircraft were processed for delivery to the National Guard. Colonel Moore is retired and lives in San Clemente, CA.

Col. Maurice M. Beech: Cocooning of B-29 aircraft with protective coating for successful preservation was accomplished under his leadership.

Col. Howard A. Moody: During this time B-29, F-13, B-17, and C-46 aircraft were removed from storage and processed for reclamation. The Korean War was the focus of attention during the latter part of his tour. Colonel Moody is retired and lives in Walnut Creek, CA.

Mr. E. B. Griffin: Mr. Griffin is deceased.

Maj. Richard P. Wollenberg, Assistant Employment Officer and Chief of the Civilian Personnel Branch, January 1942—July 1942 (not pictured): Under his leadership, the recruiting program to hire civilians for Wellston Air Depot was put into effect. At his suggestion, Mr. Karl McPherson was appointed Civilian Personnel officer for the depot by Colonel Thomas in February 1942. Major Wollenberg was reassigned to the Engineering Department.

First Lt. Alexander L. Wiener, Chief, Civilian Personnel Branch and Chief of Employment Branch, July 1942—October 1942 (not pictured).

Maj. Magnus Altmayer, Chief, Civilian Personnel Branch, October 1942—December 1943 (not pictured).

Lt. Col. Martin A. Bateman, Chief, Personnel and Training Division, February 1943—December 1943 (not pictured): Colonel Bateman left Robins for an overseas assignment.

Lt. Col. Morris Martin, Chief, Personnel and Training Division, January 1944—July 1944 (not pictured).

Lt. Col. Charles P Brown, Chief, Personnel and Training Division, July 1944—September 1944 (not pictured).

Col. James W. Mosteller, Jr., Chief, Personnel and Training Division, September 1944—March 1946 (not pictured).

Col. George W. Polk, Jr.: Colonel Polk had previously served as deputy commander and was reassigned as Special Assistant to the WRAMA Commander on May 11, 1948.

Lt. Col. John L. Grimes, Chief, Personnel and Administration, January 1949—July 1949 (not pictured).

Lt. Col. Samuel L. Moise, Chief Personnel and Administration, July 1949—January 1950: Colonel Moise left Robins to attend the Air Command and Staff School at Maxwell AFB, AL.

Mr. Karl McPherson: Mr. McPherson was tasked with staffing a large aircraft repair, overhaul and logistics support facility in a primarily agricultural region. Mr. McPherson's success in this effort coupled with subsequent related developments had the eventual effect of orienting the region toward an aerospace economy. In the 1940s, there were not enough of the skills available locally to maintain and overhaul complex aircraft systems. He developed a program to hire people throughout the South who were sent to existing depots at Middletown, PA and Fairfield, OH to learn their new trades. A later challenge was brought about by the closing of three prime depots. Approximately 3,500 new personnel with varying skills had to be assimilated into WRAMA's population. Other major changes and improvements instituted at Robins under McPherson's leadership include: the idea for a mechanized merit promotion system, the establishment of a federal credit union, the establishment of a Master SF 52 plan which reduced lapse time to fill a position vacancy by as much as 60 days, and the establishment of Toastmasters and Better Management Association groups at Robins. Mr. McPherson is deceased.

Lt. Col. Sidney B. McNair: Colonel McNair saw the need for an organization to furnish the commander and his staff with information and guidelines needed to attain the most effective management. Also, this was the initial establishment of the Comptroller in each of the military services. Colonel McNair became Base Executive in February 1952.

Lt. Col. Ralph M. Fawcett: Colonel Fawcett was transferred to Wright-Patterson AFB, Ohio.

CHAPTER TWO: *The 1950s: Organizing for the Technological Age*

Introduction

A frightened nation greeted the decade of the 1950s. In January, Alger Hiss was convicted of espionage. In February, Sen. Joseph McCarthy of Wisconsin charged that there were a number of communists still shaping foreign policy in the State Department.[1] The Korean War began on June 25, 1950. In August 1950 President Truman authorized the dismissal, in sensitive departments, of "bad security risks."[2] Differences in how to conduct the Korean War caused President Truman to relieve General MacArthur of his command on April 11, 1951.

Discontent over the drawn-out Korean War, rising prices and taxes, and corruption in Washington are usually cited as causes for the election of a former general, Dwight D. Eisenhower, as president in the election of 1952.

Throughout the decade of the 1950s, Robins Air Force Base began organizing for the technological age. The Sleeping Giant began to emerge.

With the advent of the Korean War, Robins began an expansion program equal only to that of the days of World War II. From a low of 4,161 employees in April 1947, depot population jumped to 17,697 by December 1952. There was much housing construction ($1.5 million in 1950 and $3.5 million in 1951) as well as the largest building program in history, $26.5 million, in July 1958.

Aside from the tragic tornado of April 30, 1953, which caused $2 million in real property loss, probably the biggest single development during the 1950s was decentralization. This involved transfer of certain management duties from Headquarters Air Materiel

President Dwight D. Eisenhower.

[1]The Senate Foreign Relations Committee reported that Senator McCarthy's charges were "the most nefarious campaigns of half-truths and untruths in the history of the Republic." The committee found not a single communist on his list of 205 names.

[2]By 1951 more than three million government employees had been cleared, over two thousand had resigned and 212 had been dismissed.

Command at Wright-Patterson Air Force Base, Ohio, to WRAMA during 1952 and 1953. It paved the way for WRAMA to take on the global responsibility of supply and maintenance functions.

By the end of 1954, the jet age came to WRAMA. To accommodate the modern jet aircraft Robins had to build new, longer runways.

Rapid technological changes in logistics demanded more complex, costly and changing weapon systems. And, because of the austere times, WRAMA had to produce more with less money and manpower.

Amazingly, WRAMA emerged from the 1950s with a basic organization of Materiel Management, Distribution, Procurement, Maintenance, and Base Commander which has endured for the next three decades.

The 1950s was a decade of consolidation, yet expansion with massive changes both in weapons technology and management technology. It was a time when WRAMA management faced its stiffest test—and came through with flying colors.

The Korean War

On May 20, 1950, there was a huge parade in downtown Macon, an Open House, and general jubilation for thousands of Middle Georgians, who observed the first national Armed Forces Day at Robins AFB. Music, marching, and a display of Air Force equipment pointed up the theme for the event—"Teamed for Defense." Visitors entering the industrial area at Robins walked under an arch formed by placing two B-29s side-by-side so the wings almost met. One B-29 was in active service; the other was in cocoon storage.

Thirty-six days later—on June 25, 1950, North Korea sent military forces into the Republic of South Korea in a lightning-like thrust and would later occupy almost the entire country. The same equipment seen during Open House—parachutes, radar, jet aircraft, B-29s, etc.—was soon to be prepared for use overseas. Many of the B-29s were removed from storage and processed by Robins workers for duty in Korea.

This activity was accompanied by a big buildup of manpower at Robins. Since the end of World War II, the depot had undergone large reductions in workload and manpower—from more than 20,000 to a low of 4,161 military and civilian employees by March 1947. But, depot population had stabilized at nearly 6,800 employees when the Korean War began. Of course, the Korean action changed all this and the buildup was almost a rerun of the early days of World War II.

B-29s were on display during an Open House celebration in May 1950. They later were to be put into service when the Korean War began.

B-29 Aircraft were taken out of cocoon storage and put into service for the Korean War.

Within six months, the strength had jumped to more than 9,000, and by the end of December 1952, it had climbed to 17,697.

An extensive training program naturally accompanied the hiring program for the thousands of new employees. Most of them were trained on specific operations at their work stations. This on-the-job training enabled them to begin producing in a relatively short time under normal supervision.

The earlier Armed Forces Day theme, "Teamed for Defense," became a way of life for WRAMA logisticians who began providing supplies and services to Air Force operational units in the Far East.

From the outset, speed was emphasized in resupply and the day of the propeller would wane. The jet age had arrived. There were such Air Materiel Command projects as "Hold Off" and "Speed" which demanded urgency of action. These projects had the highest priorities and absorbed much of the available manhours.

Overtime was freely used, especially until new employees could be hired and trained. Although the Directorate of Maintenance concentrated on support of Far East Air Forces, a great many jet aircraft—F-80s and F-82s—were processed for the Air National Guard, which was activated in October 1950. The ANG was called on to support the Far East Air Forces also.

Base security was strengthened and the number of guards was increased. Identification passes were changed and the industrial area was separated from the rest of the base by a chain link fence.

Through 1951, aircraft repair work increased. The advantages of the production line system, installed in May, were shown in the one-third reduction of time needed to overhaul a B-29—from twenty-four thousand manhours to sixteen thousand manhours. But there were obstacles to this line flow. The main hangar was not well suited for B-29 line repair, because the space between certain columns was less than the wingspread of a B-29. Hence, moving aircraft along the production line presented a problem. A temporary solution was to turn the landing gear so that wings were not at right angles to the line of passage. This problem was eliminated by a project which removed

One of the major jobs performed by WRAMA in the Korean War was the modification of wings on the F-84 aircraft.

A B-29 undergoes repair in a maintenance shop.

The disassembly line in Aircraft Engine Repair at Robins was busy during the Korean War.

The Sub-Assembly Unit shop of maintenance was kept busy handling engine repair work during the Korean War.

the columns. Some of the aircraft repair projects involved removal of twenty-eight B-29s from storage, reconditioning and inspecting them, removing armament and installing eight tow-target reels and power units on four B-29s.

In addition to the B-29 production line, one was set up for overhaul of F-84 aircraft. Wing modifications of twenty-five F-84s were among the projects worked on this second line.

An innovative management tool employed in maintenance during 1951 was the aircraft locator board. The board consisted of a map showing buildings and concrete aprons where aircraft repair work was being performed. The map was mounted on a ferrous backing with plexiglas patterns of aircraft made to scale. These aircraft patterns were magnetized and repositioned on the board as the aircraft they represented progressed through the various repair stations, thus showing at a glance the location and status of any particular aircraft.

Other important work in Maintenance involved installation of communications equipment at such bases as Dobbins, Hunter, and Robins in Georgia, MacDill and Patrick in Florida, and Greenville and Shaw Air Force Bases in South Carolina. These projects included a variety of telephone, radio, and radar installations.

WRAMA continued this intense level of effort until the signing of an armistice at Panmunjon, South Korea, on July 27, 1953. Base employees were justifiably proud of their contribution to the air superiority which Far East Air Forces established early in the action, and never relinquished.

Expansion—Construction

The Korean War build-up in workload and manpower was accompanied by construction activity reminiscent of the days of World War II. Additional construction was not confined to the base itself. Two months after the war began, on August 23, 1950, ground was broken for three hundred units of Wherry housing, called the "Ignico Apartments" in honor of the former WRAMA commander, Brig. Gen. Robert V. Ignico. Wherry housing was named after the late Sen. Kenneth S. Wherry (R-Neb) who introduced the bill under Title 8 to the National Housing Act on August 8, 1949. Located in Warner Robins, away from the base and bordering Davis Drive, these were brick units occupied by officers and airmen.

Some questioned locating these facilities in the city rather than on base because, by putting the Wherry houses off the main base, Robins would tie itself to a non-continguous area. However, community leaders and base officials at the time believed land on Robins should be reserved for further industrial expansion of the base. Subsequent events illustrate the wisdom of that decision. The land on which the Wherry housing would be located had been purchased by the government when Robins was founded.

The masonry apartments cost $1.5 million. Just months later, in early 1951, construction began on the second increment of two hundred additional units.

Then on March 7, 1951, Congress announced the first big construction on Robins since World War II—a $3.5 million allotment. Among the items funded was a million dollar project to expand the aircraft parking apron to accommodate heavier aircraft.

Other important construction items included in the allocation were a hazardous test building, a powered equipment repair shop and the first nine holes of a new golf course (the second nine holes were completed in August 1968) as well as several rehabilitation projects.

Construction continued at a rapid pace on base during 1952. In addition to the new hazardous test building, construction of a new paint house, a vehicle repair building, east and west extensions to the engine repair building, a new communications building, and two new warehouses was under way. And, six of the engine test cells were being modernized.

Following the buildup in 1952, Robins AFB was declared a permanent installation, after a decade of civic effort to have it so designated. This encouraged private construction and permanent residence in the City of Warner Robins.

The devastating tornado of April 30, 1953, cost the base $2 million in real property loss. However, it also occasioned an immediate surge in building, most notably the headquarters office building, building 215, which was completed and occupied by early 1955.

In February 1955, Robins began construction of a modern runway which would accommodate any land based jet aircraft then in production. Later that same year the federal government acquired title to three tracts of land aggregating 211.45 acres for completion of a new runway extension.

A new streamlined fire and crash station opened on December 19, 1955, replacing a temporary type structure in use since 1942. Three warehouses at a cost of $3,415,453, an addition to the heating plant, and the new headquarters office building were also among the construction for that year.

One of two new warehouses, covering twenty-seven acres of ground under roof, which were completed in the fall of 1954.

This Wherry Housing, known as Ignico Apartments, housed military members in Warner Robins.

Building began on a new aircraft parking apron at Robins in March 1951. It provided outside work space and parking area for aircraft. Almost completed in the center is a concrete aircraft nose dock, or shelter, where aircraft were worked on. C-124s of the 7th Logistics Support Squadron are shown.

To exploit automation, a giant mechanized handling system went into operation in August 1956. The system moved Air Force supplies from commercial carriers through receiving, inspection, corrosion control and packaging, and dispatched them for delivery to the storage area. An in-the-floor drag chain conveyor connected the system with some twenty-seven acres of warehouse space. The drag chain, about a mile and a half long, normally traveled at the rate of one hundred feet per minute and was believed to be the longest in the Air Force. Powered conveyors automatically sorted and routed packages through the various phases of shipping and receiving. By simply pressing a button, a package was moved to its station by an electronically-controlled dispatch unit. These units moved and separated boxes weighing up to four hundred pounds each at the rate of six a minute.

Meanwhile, in 1956 the base let some $1.5 million in bids for new construction, including $765,814 for a parking apron. Another contract for $570,499 was awarded for construction of two modern dormitories in the troop area.

This increase in defense expenditures was accompanied by a build-up in military personnel. Thus, not only operational facilities, but housing and personnel support buildings were required as well.

In April 1958, Robins embarked on the largest construction program since the completion of the base in 1942. Contracts for construction projects approxi-

Building 350 was one of three supply warehouses which added 800,000 square feet in covered area to the WRAMA warehousing capability. The warehouses were four hundred feet wide. One was eight hundred feet long and two were six hundred feet long.

A new headquarters building for WRAMA was among construction projects in late 1955.

mating $26.5 million were awarded by July 1, 1958, and projected for completion by January 1960.

It was evident from this construction program that the Air Force planned to expand greatly the scope of operations at Robins. It was not until July 1958, however, that the public was informed that a unit of the Strategic Air Command (SAC), the 4137th Strategic Wing, was to be stationed here.

Meanwhile, the construction program had begun. The area to the east of the runway was chosen for SAC operations. Here a heavy-duty concrete apron was to be built for which underground air, water, fuel, and power facilities necessary for servicing planes would be installed. Congress funded the sum of $4,473,900 for apron construction in March 1958.

In late 1957 the Air Force had scheduled construction of 273 units of Capehart family housing (named for the late Sen. Herman E. Capehart, [R-Ind.] who sponsored the bill).

Located in the southeast corner of Robins, these were the first family housing units for military personnel to be located on base since the 1940s when the commander's quarters, C Circle, and Officers' Circle had been built. The 273 units of Capehart housing were finished in June 1959. A year later, in June 1960, an additional 150 units of Capehart housing were completed on Robins. In all, Robins had 423 units of housing authorized under the Capehart housing project.

In February 1958, the base assumed control of five hundred Wherry housing units in Warner Robins from the Fickling and Walker Rental Agency which had built them in 1951 (three hundred units) and 1953 (two hundred). These apartments locally referred to as "Ignico," were modified by the Air Force shortly after they were acquired. After eighty of the one-bedroom units were converted into forty two-bedroom units, there were 460 units left. A contract

Two modern dormitories were built in the troop area at Robins in 1956.

Under construction in late 1957 were 273 units of Capehart family housing similar to these officers' units located in the Forest Park area of Robins.

Another 150 units of Capehart housing were being built in June 1960 like these enlisted personnel homes in Crestview.

Five hundred units of Wherry housing, such as this house in Chevron Village, were transferred from Fickling and Walker Rental Agency in Warner Robins to Robins AFB in February 1958.

for Wherry modernization was awarded in late 1958. Work on the brick tile houses involved general repairs, heating system improvement and driveway paving. The houses were popular among field grade officer (majors and above) occupants. Thus the Wherry housing units together with Capehart housing provided Robins Air Force Base with housing for fifty-five percent of its officers and upper-grade airmen.

Of the original 1,058 units of Lanham Act "Zeigler" housing, 126 units were destroyed by the 1953 tornado, 840 were disposed of between June 30, 1960 and July 1, 1965, and the remainder were converted from one-bedroom into efficiency units.

In addition to the Capehart housing building and the Wherry housing modernization, other construction projects provided in the $26.5 million construction program begun in 1958 included an aircraft maintenance hangar, four airmen's dormitories, three officers' quarters buildings, a squadron headquarters, and an operations building for the 7th Logistics Support Squadron (LSS) as well as a parking apron in front of the 7th LSS area, a new communications building which serviced base units including the 1803rd Army Airways Communications System (AACS) Group, the 1926th AACS Squadron, the 1852 AACS Flight, Detachment 3, 1889th AACS Installation and Maintenance Squadron, 4th Communications Construction Squadron, a new inward-outward dial system, and a new base gymnasium.

This type of Zeigler (Lanham Act) Housing left the scene in July 1965 when the last of 406 units built in 1943 were disposed of.

Extending almost the entire length of a huge aircraft maintenance hangar at Robins AFB was a two-story addition which increased the WRAMA shop space by more than 128,000 square feet.

NCO Wives' Club

After having operated as a women's auxiliary from 1951 to 1955, the Non-Commissioned Officers' Wives' Club was formed on March 7, 1955, at a tea attended by 125 wives and guests.

The club has always been active in the mainstream of life at Robins AFB and in the civilian community by providing volunteer work and many donations to different organizations on and off base. It has representatives on the Commissary, Base Exchange, Child Care Center, Hospital, NCO Advisory, Family Services, and many other boards in the community.

In addition to giving parties for senior citizens at the Hallmark Nursing Home in Warner Robins and donating money to the Air Force Aid Society, the club has helped many community families in emergency situations. The club also awards three five hundred dollar scholarships each year to college-bound students.

Two of four dormitories which were completed in late 1959 gave Robins six dorms to house its airmen power.

Decentralization

The 1950s can be called the Korean War years period of expansion. But the single development during the 1950s most important to Robins' future was the transfer of certain logistics management responsibility from the Headquarters Air Materiel Command (HQ AMC) at Wright-Patterson Air Force Base, Ohio, to all the Air Materiel Areas, including the Warner Robins Air Materiel Area at Robins Air Force Base.

This map shows nine AMAs and four special activities in the command in 1952.

This development, known as decentralization was caused by a number of things. The command realized it could not efficiently conduct its many operations from one main location. The increasing size and complexity of operations at the Ohio headquarters and the overcrowded facilities at Wright-Patterson AFB were immediate factors in the decision to decentralize.

Among long-range considerations were the strategic vulnerability of the headquarters and the need to establish an organizational structure which could expand. With the existing setup, the Air Force logistics system could be crippled by an attack on Wright-Patterson AFB. Also, the headquarters had just about reached its capacity in the area and, in case of mobilization expansion, would be hard put to accommodate other functions within its existing framework. The move would also permit the headquarters to concentrate on policy, planning, and management control, its principal responsibilities.

Prior to decentralization, WRAMA had received instructions from the command headquarters in Ohio. The managers there directed the storage, shipment, or issue of items and the repair and overhaul of airplanes and components in the Robins shops. Beginning in 1952, the management function was decentralized to the various depots and for the first time, the brain and muscle power at Robins could work together. The change paved the way for an international flavor at Robins, for out of "decentralization" emerged the concept of "prime" responsibility. WRAMA now had prime control over assigned aircraft and commodities worldwide. As a result, the significance of WRAMA to the geographical area diminished and its worldwide significance increased, as Air Force units began to look to the prime managers for support. This awareness of WRAMA's global mission was a factor which caused the city of Warner Robins to take on the name "International City." Another factor was the diverse population mix associated with Robins AFB.

There were two types of global responsibilities involved: (1) responsibility for a given "commodity," such as propellers, tires or electronic warfare equipment; and (2) "system" responsibility wherein a depot was responsible for integrating all of the commodities to provide a functioning aircraft system.

Involved mainly in the decentralization were the WRAMA functions in supply which bore the brunt of the program with seventy percent of the two thousand persons transferred to the various centers from Wright-Patterson AFB coming from and going to Supply.

Billy C. Humphrey, radar repairer, aligns and calibrates a computer used on the fire control system in fighter-interceptors and bombers.

Within the structure at Robins, a parallel decentralization move of sorts was made on August 1, 1953, with the formation of the base commander organization. Before then, except for a short period during World War II, the three organizations—headquarters, depot, and base—were integrated and operated as a single activity.

In the early 1950s, the command's philosophy was that the area commanders should devote maximum attention to primary mission responsibilities and delegate responsibility for support functions to the appropriate members of their staffs. Thus the base commander organization was born to free the commander of administrative matters so that he could devote his time to the larger and more pressing problems of depot and area operations. The base commander organization functioned as the 2853rd Air Base Wing.

Two months after decentralization—on October 1, 1952—HQ AMC made an important change in its procurement organization. In addition to establishing a procurement directorate, the command discontinued the former procurement districts. What had been

Decentralization . . . Challenge to Management

The Air Force has a commendable record of employee relations, particularly when transfers of functions and responsibilities to other geographical areas has been determined to be wise. It was the philosophy of the Air Material Command at the time to let people assigned to the headquarters in Dayton, Ohio, volunteer to go to the Air Materiel Area they most preferred. Allowing potential transferees to choose locations encouraged retention of skills. Fewer chose Warner Robins, the most rural choice, than any other location.

WRAMA management thus faced a tremendous challenge to hire and train people quickly to carry on the operation of a mission which was itself newly assigned. It is a tribute to WRAMA management and those new people that mission performance did not falter.

When the specialized depots at Topeka, Kansas, and Mansfield or Shelby, Ohio, and some other places were closed, the percentage of people functionally transferred to WRAMA began to increase.

During the 1950s, prior to the closing of the depots at Middletown, Pennsylvania, and Mobile, Alabama, and other installations in 1964-1965, WRAMA had absorbed more work from Air Force installations phasing down than any other AMA.

The A-3A bombing fire control system of the B-52 was overhauled in the WRAMA shops. Left to Right: D. C. Crawford, L. J. Kellum, W. T. Williams, J. R. Green, A. L. Thompson, and T. A. Barley.

named air regional offices were renamed air procurement districts and placed under the jurisdiction of the AMAs. The Air Force Plant Representative (AFPR) offices were also placed under the jurisdiction of the AMAs.

Assigned to WRAMA at equivalent directorate level were the Atlanta Air Procurement District; AFPR Office, Glenn L. Martin Company, Baltimore, Maryland; and AFPR, Curtiss-Wright Corporation, Caldwell, New Jersey.

Prime depot responsibility as well as geography determined the assignment of the procurement activities among the AMAs.

The reassignment of the procurement activities was in line with Air Force policies at the time to eliminate intermediate headquarters and duplication of effort where possible and to make manpower available for higher priority projects. The former air procurement districts had been intermediate headquarters.

Upon completion of decentralization in late 1953, WRAMA assumed global responsibility for supply and maintenance of its assigned systems and commodities.

Vehicle Management

During the early 1950s, the Air Force still had a sizeable number of specialized depots in Maywood, California; Topeka, Kansas; and in Memphis, Tennessee. Except for these and others being built, the specialized depots had phased out after the war. Others were operated throughout the United States in leased facilities.

The last functional responsibility decentralized from HQ AMC in 1953 was the family of vehicles. But their initial assignment to WRAMA lasted only until the summer of 1954.

Early in 1954 great management strides were being made in the identity, serialization, and control of vehicular end items by WRAMA. Beginning was the philosophy of establishing local purchase functions to support vehicles so the Air Force did not have to maintain massive warehouses of stock of such things as piston rings, fan belts, etc., which could be bought locally.

This philosophy took root at WRAMA in 1954. As a result of overall growth and evolution in the Department of Defense's logistics operation, various functions were peeled away to the Defense Logistics Agency (DLA), General Services Administration (GSA), etc., as well as some local purchase authorization to foreign countries.

This local purchase philosophy was advanced as early as 1953 and in November 1953, Mr. Cotton, Assistant to the Director of Supply and Services, went to Japan to set up the first local purchase program ever by American forces with a foreign country. Because of the tremendous difficulty with Japanese customs and the involvement with the Japanese occupation by the American forces at this time, the program failed. However, things learned and experience gained from the trip did begin to emerge during the war in Southeast Asia as a viable alternative which later became an operating principal.

In the summer of 1954 it was decided to relocate the vehicle management responsibility from WRAMA to Memphis, Tennessee, since the Air Force and the AMAs were seen to be more aircraft and systems management oriented than commodity management oriented.

Memphis continued to manage vehicles until 1962, when it was decided to close the specialized depot. The function was retransferred, at that time, back to WRAMA.

Vehicle-launched TM-61

Army helicopters serviced at Robins by refueling vehicle.

This photograph, taken from barracks number 1510 at Robins AFB, shows the tornado moving through the base.

Tornado—1953

What a difference twelve minutes makes!

On April 30, 1953, at 5:12 p.m., a boiling, black tornado ripped across Middle Georgia's countryside and into the city of Warner Robins and Robins Air Force Base.

The twister left in its wake 18 dead, 350 injured, 1,000 homeless, and $10 million in damage. It stands out as the most catastrophic event in the annals of Robins AFB.

The tornado, which caught everyone unaware, cut a swath from three hundred to five hundred yards wide through the city of Warner Robins and Robins AFB, then crossed the Ocmulgee River and continued for about twenty miles to the vicinity of Irwinton. It went from a southwesterly to a northeasterly direction.

Except for two fortuitous circumstances it could have been worse. First, the storm hit the base twelve minutes past five o'clock and most employees on the day shift had left. Twelve minutes earlier and it would have caught many more people leaving the gate or at their jobs. For instance, an administration building which ordinarily employed one hundred people was smashed. They had all gone home. Second, much of the tornado's narrow path through the base was over the open storage area. It missed the Post Exchange by almost thirty yards but it leveled the nearby Post Office. There were some fifty people in the PX but, luckily, it was not in the twister's path. Casualities would have been higher if its course had been a short distance to either side.

Of the eighteen fatalities, fourteen occurred in the area of the city and the base and four were in communities east of the Ocmulgee River. Most of the dead were women and children. For Robins, there was one death and twenty-six disabling injuries. The base's lone fatality was an officer's wife on her way to pick up her husband from work. She and her two children arrived at the gate just as the tornado hit the base. An eyewitness recalled that she was killed by a post which was driven through her car. Her children were not found for several hours. A concerned person (never identified) had found them and taken them to a Macon hospital for treatment. They were not seriously hurt.

Many tales were recounted after the storm. One Twiggs County deputy sheriff picked up two small children in a cotton field after they were blown two blocks from where their home stood—or had once stood. The children were kept alive through artificial respiration.

Another eyewitness said he heard some noise, but "thought it was a jet aircraft," and the last thing he remembered was closing the door in his Zeigler Apartment as "bricks began falling on me."

Others told of seeing bodies hurtling through the air and of persons huddled in ditches to escape the savage fury of the cone-shaped twister.

An Air Force captain who was in Germany during World War II and had been in Hawaii during Pearl Harbor, observed: "I've seen places bombed that were not this badly damaged." His wife, a German, said, "It looks like a bombing raid."

One Robins airman, A1C Robert Smothers, according to newspaper accounts, was in the mess hall when the catastrophe struck. He yelled "tornado!" as the storm hit. Smothers dashed outside and pulled a bloody and presumably badly-hurt girl out of a car which had been turned upside down. He saw a civilian nearby under a pile of heavy timbers and believed he was dead. He pulled the timbers off the man. He was alive.

The late Joe Parham, editor of *The Macon News,* said, "The caprice and might of the storm were self-evident. A two-by-four piece of timber had been driven through a metal car door. The ends of uprooted pipes stood sentinel over wrecked houses like ghostly skeleton fingers."

Robins played a key humanitarian role during and after the storm. The Base Domestic Emergency Plan was put into operation. Under the plan, teams helped with hospitalizing the injured, searching for victims, repairing utilities, and distributing medicine, food, and bedding. Robins also provided other emergency supplies. The base hospital became the medical center for treating the injured.

By noon the following day, Friday, electric power and water were restored to all except the areas which were hit by the tornado. However, cleanup operations continued for several months.

Destroyed were 126 units of government-owned South Zeigler housing and 182 Ignico Apartments built by Fickling and Walker, Inc., of Macon. Loss to real property on the base exceeded $2 million. In addition, there were heavy losses of materiel and supplies.

Wreckage was scattered over an open area of Robins. Base officials discuss the damage in upper right.

The South Zeigler section of Warner Robins was hard hit. This is view looking northwest.

The south and open area of Robins AFB following the tornado.

Half of Building 1350 was left standing after the twister hit Robins.

The path of the tornado, looking northwest through the industrial area of Robins Air Force Base.

This is what remained of Building 1154.

Desks were all that remained of Building 1179.

What remained of the Air Installation Office.

The roof and front section of Building 115 were ripped off by the twister.

The roof collapsed on Building 1142.

Military guards maintain order in the Ignico housing area.

Zeigler housing occupants check their houses for any valuables left after the twister.

Refrigerators and bedding were piled up in Zeigler housing.

The entire side of the building leading into the kitchen of this Ignico apartment was ripped off by the tornado.

A Zeigler apartment occupant, head still bandaged from having been injured in the tornado, surveys the damage after it hit the apartments.

The trememaous force of the tornado whipped this car from its driveway and hurled it against a once stately Georgia tree.

Cars were demolished by the tornado in the front of Ignico apartments.

Little Caron Dodds, daughter of MSgt. Harold Dodds of the 6th Weather Squadron, looks undisturbed at the surroundings of her crumbled home.

A mother and her baby check over the remains of their Zeigler apartment.

Mr. and Mrs. Julian Crutchfield and son, Billy, look down at the spot where they crouched when the twister struck their home, then bombarded them with wall fragments and splintered lumber.

Maintenance workers in Aircraft Repair overhaul F-86s which are lined up in Building 1107. On the left are lines of T-33s and B-57s being repaired.

Jets and Missiles

After three years of combat in Korea, the combatants ended up only a few miles—give or take a little—from where they started on June 25, 1950. (The armistice agreement was signed July 27, 1953, at 10 a.m. at Panmunjom.)

Just following the truce, the world was stunned by the announcement in the summer of 1953 that Russia had successfully tested its first hydrogen, or thermonuclear, bomb. The word came only four years after the Soviets had tested an atomic bomb. Four years later, Russia boasted that it had test-flown an intercontinental ballistic missile capable of carrying A- or H-Bombs from Russia to any point in the United States.

President Eisenhower placed new emphasis on air power and the principle of "massive retaliation" as the means to deter Communist aggression. The Air Force budget became larger than that of the other armed services. Everyone realized that the chief threat to the country came from Russian nuclear capability.

The Korean War and related events drove personnel strength of the Strategic Air Command from seventy thousand to almost two hundred thousand—larger than the whole U.S. Army in 1939—and SAC aircraft from one thousand to more than three thousand.

Throughout the Korean War, WRAMA repaired the F-80, F-84, and F-86 jet aircraft. By 1954, WRAMA no longer worked on these fighter aircraft which were being replaced by the century-series jet fighters. Robins AFB was expecting a long and gradual transition from planes to missiles and had begun to prepare for the change in the spring of 1954.

WRAMA had prime system management responsibilities for the TM-61 Matador. This pilotless jet bomber was a surface-to-surface missile, the first of its kind in the USAF inventory. The Matadors were assigned to the North Atlantic Treaty Organization (NATO) defense forces in Germany and the first squadron was deployed to Europe on March 9, 1954.

The jet age was upon us by the end of 1954. All fighter aircraft in the regular Air Force were jet-propelled and more than fifty of the combat wings were fully jet-equipped. The old propeller-driven P-51s, repaired at Robins during the war, had faded from sight. The last B-29 was retired on November 1, 1954, having completed an operational life cycle of ten years.

The Air Force announced in 1954 that a new jet tanker, the KC-135, was placed on contract and in August 1954 the eight-jet B-52 Stratofortress intercontinental bomber entered the Air Force inventory. Unlike previous aircraft which were retired when new technology made them obsolete, these aircraft were destined to be a formidable team through the present day, twenty-eight years later and counting. Both were to roost at Robins AFB with the 4137th Strategic Wing (now the 19th Bombardment Wing) after it was activated on July 1, 1959. SAC's move to Robins Air Force Base was a part of a plan to disperse its combat strength over a larger number of bases. The speed with which SAC could launch its bombers might well mean the difference between success and failure. More bases would certainly improve the ability to get the fleet airborne.

Meanwhile, Air Force production lines were increased about thirty-five percent in 1955 and the F-101 and F-104 supersonic fighters were put on an accelerated program.

WRAMA entered the jet era as a System Manager when it was announced late in 1954 that it had assumed systems support responsibility for the new medium jet bomber, the B-57 Canberra. WRAMA technicians made periodic temporary duty trips to resolve supply, support equipment, and maintenance problems with the B-57.

The old propeller-driven P-51s were phased out of the Air Force inventory in 1951. They had been repaired at Robins during World War II.

Robins AFB Maintenance workers repaired the F-80 jet aircraft during the Korean War.

The TM-61 Matador missile was assigned to WRAMA for logistical support in 1954.

At Lockheed-Georgia Company in Marietta, Georgia, the first production model of the C-130A Hercules turbo-prop transport came off the lines in April 1955. Three years later—on January 1, 1958—prime responsibility for the aircraft was transferred from Sacramento AMA, California, to WRAMA; it is still managed and repaired here.

The impetus given the ballistic missile and aircraft programs in 1954 began to pay off. Jet bombers took over in 1957 as the principal weapon for deterrence; missile development was further accelerated; technological breakthroughs occurred almost daily.

As a follow-on to the TM-61 Matador, WRAMA assumed management of the TM-76 Mace missile. In keeping with the new logistic concept for missiles, HQ AMC directed WRAMA to make plans for establishing a capability for "in-house" maintenance support of the TM-76 Mace.

The Canberra B-57, a WRAMA prime aircraft, is being transferred into a hangar to undergo repair at Robins.

WRAMA managed the TM-76 Mace missile as a follow-on to the Matador.

In addition to missile support, Robins was involved in a mammoth Air Force project known as "High Flight." Under it, the first F-100C supersonic fighter landed at Robins on October 27, 1955, for inspection, winterization, and flight test prior to flying overseas. This project, which begn in March 1953, consisted of delivering jet aircraft—bombers, fighters, and trainers—to U. S. Air Forces in Europe (USAFE) and Mutual Defense Assistance Program recipients in Europe. These countries included France, the United Kingdom, Portugal, Spain, Italy, Germany, the Netherlands, Belgium and Greece.

To accommodate all these modern jet aircraft, Robins had to build new, longer runways. Current runways had been built in 1942 for World War II aircraft. In 1955 a new runway was designed to accommodate any land-based aircraft in production or on the drawing boards. All three of the old runways were becoming obsolete.

WRAMA maintenance repaired the avionics systems of the F-104 supersonic fighter.

A WRAMA managed C-130A just before it was flight tested at Robins. On the right is G. T. Lowery of Flight Test Section along with the flight test crew.

The F-100C underwent modification prior to flying overseas under Project High Flight in March 1953.

All of this new technology pointed up one fact—the systems were becoming increasingly complex. This was borne out by the increasing number of "black boxes" in a system. (A black box was a self-contained unit, designed to do a special job, having a specialized circuit, and usually being beyond repair capability except at a specialized depot. It was characteristically black.) For instance, at Robins Air Force Base, the E-1 Fire Control System had thirty-six black boxes and MA-2 Bombing System had 204 black boxes. (The E-1 system applied to the search radar and fire control systems of the F-94 while the MA-2 was a computer system which provided the B-47 and B-52 aircraft with data on when and where to drop bombs.)

As the decade closed, on July 1, 1959, it was announced that WRAMA would assume management responsibility for the C-47, C-54, C-117, and C-118 aircraft, soon prompting Robinites to call their AMA "the cargo aircraft center of the Air Force."

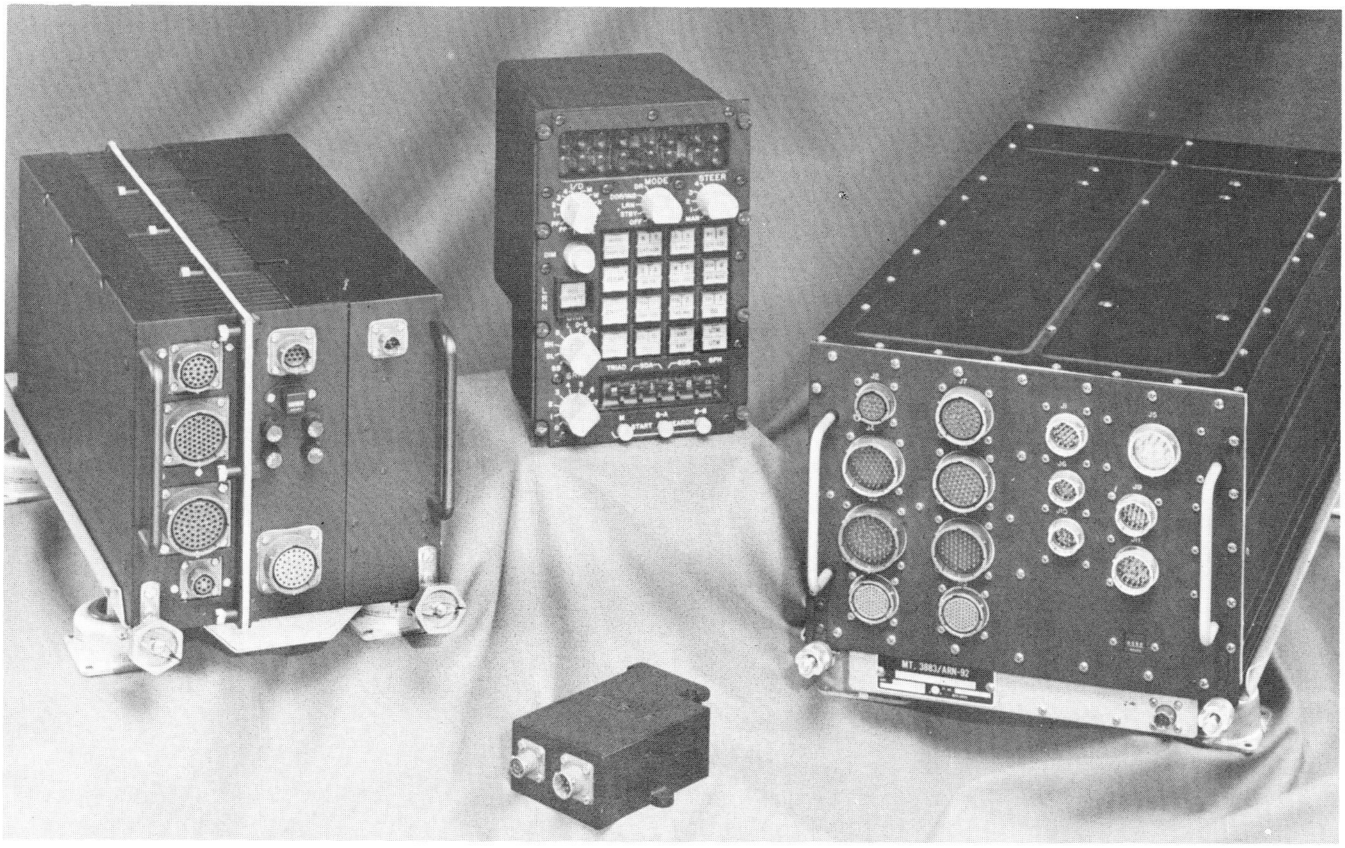

There were an increasing number of "black boxes" managed by WRAMA in the 1950s.

WRAMA assumed management responsibility for the C-54 in July 1959.

WRAMA managed the C-118 in 1959.

Three aircraft associated with WRAMA are shown in this 1949 photo: the C-47 at the top, the C-119 in the middle (both were later assigned for management), and the B-29 at the bottom which was modified in WRAMA shops.

WRAMA managed the C-140, but only a few flew in to Robins.

Mechanization

The removal of the specific pressures of the Korean War permitted the administration of President Eisenhower to develop more rapidly its general defense and foreign policies. In part they sprang from a strategic calculation—that there was no longer any one year in which the danger of war was greatest but that policies must be laid down to take care of a number of years, any one of which might be critical.

On January 12, 1954, Secretary of State John Foster Dulles stated in a speech before the Council on Foreign Relations that the administration was aiming for "a maximum deterrent" of aggression at "a bearable cost." To achieve this, it was going to deemphasize "local defense" and rely more on "the deterrent of massive retaliatory power . . . a great capacity to retaliate, instantly, by means and at times of our own choosing."

In line with the policy of massive retaliation came a "new look" at the defense budget by the Joint Chiefs of Staff which trimmed $2.3 billion in expenditures and $5.2 billion in defense spending authority from the Truman proposals for FY 1954.

It was still WRAMA's mission to support the weapons and equipment, to replace and repair them, and to sustain using commands in their theaters of operation. But the means of carrying out our logistics mission were changing. Logistical support had to be tailored to advanced weapons and force structures, different concepts of warfare, newer modes of transportation, and communications.

No longer could we depend so completely on mass logistics, huge stockpiles, and large depot industrial facilities to meet the needs of modern global warfare. Now we would need a more precise, specialized type of logistics, set for individual needs of the more complex, costly, and changing weapon systems. And WRAMA had to produce more with less money and manpower and do it faster. This challenge was compounded by rapid technological changes in logistics tools. Technology was available in the post-war period for quicker collection and flow of management and operating data.

The Air Materiel Command had begun its first mechanization program in late 1948. It mechanized its property accounts with IBM punch cards. By June 30, 1950, WRAMA completed the first phase of the mechanized property accounting project.

Under Project WISE, the Maintenance Directorate put into effect a more complete mechanization of the maintenance cost system in the Engine Repair facility.

The Directorate of Supply installed a moving belt conveyor for packages in 1951.

Supply rebuilt its Parcel Post Packing Line and installed a moving belt conveyor for packages, put into operation five sets of Rol-A-Lifts to unload freight cars, and built a system of electric hoists to lift and move heavy items in the Warehouse Section during 1951.

Now that WRAMA had gotten a taste of mechanization, a move began to improve procedures and facilities. For example, a "pony" teletype circuit for mechanically processing overseas requisitions saved time and money in the Supply Directorate.

A pneumatic tube message system, still intact although not operating today, was an innovation in October 1952 in the Directorate of Supply and Services. A trench was dug between the headquarters building (Building 220) and the supply offices, for the message tubes had to be placed underground. Requisitions flowed from the central point of receipt by tubes to the stock control clerks; they put the shipping documents in the tubes and sent them to the warehouse or storage section for selection of materiel and subsequent shipment.

One incident was recounted by an employee in Central Receiving. Employees were paid by check which in those days were handed out to each person at work. These checks were routinely sent through the tube, but this time they got stuck in it. A repair crew had to apply a greater amount of air pressure through the tube to force these checks through before the terrible prospect of a payless Friday was laid to rest. Use of the tube message system was also laid to rest by the late 1950s.

WRAMA was a pioneer in applying mechanization to procurement work. In 1956 WRAMA combined the IBM cardotype accounting machine and special slave typewriters to perform seventy percent of the typing of invitations for bids and requests for proposals.

WRAMA received its first Univac computer in middle 1954. It received two IBM 650 electronic data processing machines and a Univac file computer in April 1958 and received a larger Univac computer later that year. We were moving from mechanization to early automation.

More data automation occurred on July 1, 1957, when WRAMA converted local fund accounting from the electric accounting machine equipment to the IBM 650 computer. The IBM 650 was a marvel. It integrated appropriations accounting with general ledger accounting; it made decisions and told why.

WRAMA was also a pioneer in the application of statistical quality control techniques to clerical areas. In August 1955, preparations were made to convert from a punch-card system to a mark-sense system in mechanized property accounting. A new statistical quality control system applied to the mark-sense process was highly successful—errors in the Inventory Accounting (Monetary) System dropped from more than ten percent to less than one percent.

By 1962, the command had a rental budget of more than $34 million annually for various types of data processing equipment. In order to take fullest advantage of automation, systems had to be devised to specify and regulate the flow of raw data, to determine the management data to be produced, and to assure that the data generated in one system could be used in a related system.

Responsibility for design and development of automatic data systems was first assigned to the various air material areas. Increasing sophistication of equipment, the complexity of requirements, and the difficulty of making systems compatible, drove AMC to centralize its data development in middle 1960 at the Headquarters.

Management Innovations

Supply Management

Numerous management innovations were introduced in the 1950s as WRAMA shifted its logistics responsibilities, in step with technology changes. Until well after World War II, AMC operated its supply installations, such as Robins Air Force Base, on a geographical-general supply depot basis. WRAMA would stock a range of supplies for the use of bases within its geographical area.

Need for controls over materiel, which became increasingly complex, had made general depot operation impractical and too costly. Therefore, the command created the Two-Zone System in 1948.

In March 1950, the Air Materiel Command modified its Two Zone System. It designated one of the two zonal depots for each class making it responsible for distributing and expediting all contracts pertaining to the property class initiated at Headquarters AMC. The designation of such a depot became a "prime depot."

Decentralization changed the operating procedures set up under the Two Zone System, allowing other zonal depots to submit requirements through the prime depot instead of directly to AMC. The change improved service to the operating bases and relieved an excessive workload at HQ AMC.

The Two Zone System remained in operation until 1956, when the opposite depot was eliminated and the prime depot became designated as the lead air materiel area. Stocks at opposite depots were reduced by attrition. Assignment of responsibility was on an item or commodity basis; that is, all items in a Federal Supply Catalog group were assigned to one AMA for storage and/or management control.

HQ AMC made a break with the purely commodity management of Air Force stocks in 1958. Management of part of the stocks pertaining to a weapon system (usually peculiar to that system, i.e., having only that single application) was placed under a single Logistic Support Manager, who operated under the weapon system concept. The manager, known as System Support Manager after 1960, managed those peculiar stocks required for the operation of a weapon. Thus, the user of that system enjoyed the convenience of having a single source for supply for many of the supporting parts that heretofore had been managed by a variety of commodity or Federal Class Managers.

Faster communication, accounting, and transportation systems had lessened the need for actual stockpile possession of all stocks at the location of the commodity or system manager. Selective management principles had led to the growth of a myriad of support concepts and their ensuing supply programs.

Supply workers make a quality check on material in the Receiving area.

Inbound material is checked in Central Receiving on one of the first conveyor lines installed in the Directorate of Supply.

Workers in the shipping complex of Supply's Transportation Operations Division apply shipping labels to crates.

Better Management Association

Upgrading skills to keep up with technology was a major task of Robins Air Force Base managers during the 1950s.

To that end, a Better Management Association was formed in 1956 as a Nonappropriated Fund activity. It was originally named the Robins AFB Management Association.

In 1957 it became a private organization, chartered by the state of Georgia.

Currently, there are four groups under the parent BMA organization with a total of three hundred members— Group A, Policy and Command Management; Group B, Quality Control; Group D, Executive Development and Leadership Ability; and Group H, Engineering.

Senator Herman Talmadge (D. GA) and a group of BMA members in the Base Restaurant after a BMA function in February 1967. Left to right: Fred Houser, William Gilligan, Johnny Deep (BMA President), Cecil Jones, Sen. Talmadge, H. L. Everett, John Vassas, Margaret Hatcher, George Cornish, and Joe Molony.

Among the variety of these new support concepts was a new type of prestocking for new, high priority weapon systems such as the C-130, B-57 and the TM-61 Matador and TM-76 Mace at Robins AFB. This was done in order to have a full range of operational support requirements segregated and reserved for a given system.

Later, control was centered at the System Manager AMA and storage was determined by strategic need rather than physical proximity. As rapid communications and transportation systems became progressively more efficient in the 1960s, need for this kind of specialized segregated storage diminished.

During the 1950s, WRAMA made great strides in improving logistics transportation as well as communications and data processing. It made even greater use of aircraft for priority cargo.

Under Project Skyway in late 1951, WRAMA had three C-46s assigned here by Headquarters Air Materiel Command in a service test to determine if it wasn't cheaper to fly materiel in a "resupply by air" procedure than it was to buy spares to fill the longer pipeline associated with other resupply modes. After the test was over in 1952, it was indeed found cheaper to fly materiel to bases than to send it by truck or rail. Not only was the in-transit time a lot shorter but materiel handling was also reduced. Also, a more far-reaching affect was the elimination of prepositioned stocks overseas by depending on direct aerial resupply. The conclusion reached was that the test, which heavily influenced logistics actions, paved the way for later overseas closure actions. This test proved to be a forerunner to the LOGAIR system which AMC introduced in April 1954.

Under LOGAIR, commercial airlines contracted to move cargo among depots and bases in the Zone of the Interior. This system greatly reduced pipeline requirements and gave new elements of mobility to logistics.

A forklift was used to move items around in warehouses.

Packagers in the Directorate of Supply display the latest pakaging techniques. Left to Right: Ed Fountain, Jim Connors, Noble D. (Dee) Teal, and Gene Pierce check the packages.

Material was unloaded in the early 1950s by forklift at the Air Freight Terminal.

To keep in step with the increased scope and importance of these supply and transportation innovations, WRAMA reorganized its Supply Directorate three times—in September 1950, in July 1952, and again in February 1956.

In September 1950, the most important changes were designed to accomplish more efficiently the prime and zone phases of the mission. With this in view, the Stock Control Section was organized so that property classes were grouped more on a basis of whether the supply responsibility was prime, zone, area, or base rather than on commodity similarity.

The organizational directive of the Directorate of

Supply and Services was revised in July 1952 as a result of the decentralization program. The components transferred from HQ AMC were grouped under a Deputy for Requirements and Distribution, while the name Materiel Control Division was used for the former Requirements and Distribution Division.

The February 1956 reorganization involved the development of the Federal Supply Classification System to replace the separate catalog systems of the Air Force, Army, and the Navy. Under the Federal Supply Class system, a primary commodity area was termed a "group" and a subdivision of a group was termed a "class." A four digit coding structure was used to show commodity areas, the first two digits denoting the group and the last two the class. Individual items were identified by a seven-digit number following the class code.

The Air Force's method of conversion was to implement a federal group and transfer to it the applicable items from the old Air Force stock classes, wherever they were managed. The Air Force classes remained active as long as they had any items.

WRAMA was little affected by conversion to federal cataloging until March 1955, when the pace of conversion increased and migration of prime items from WRAMA began. Final implementation and operational refinement was completed by the middle of 1956.

During this period of rapid management system evolution, we never lost sight of our primary mission, and, while accommodating to the changing ways of doing business, the customer was always satisfied.

Directorate of Supply leaders in 1950. Left to Right, 1st Row: *Lt. Col. Johnson, A. H. Cotton, Col. L. B. Dietrickson, Maj. Hoban, E. W. Smith, and Lt. Smith.* 2nd Row: *Paul Fain, Maj. Hardy, D. E. Williams, Capt. Lorenz, and K. C. McPherson.* 3rd Row: *Lt. Shelton, Bill Harrison, Capt. Ford, W. G. Jenkins, Capt. Kay, E. M. Waters, and Capt. Lowe.*

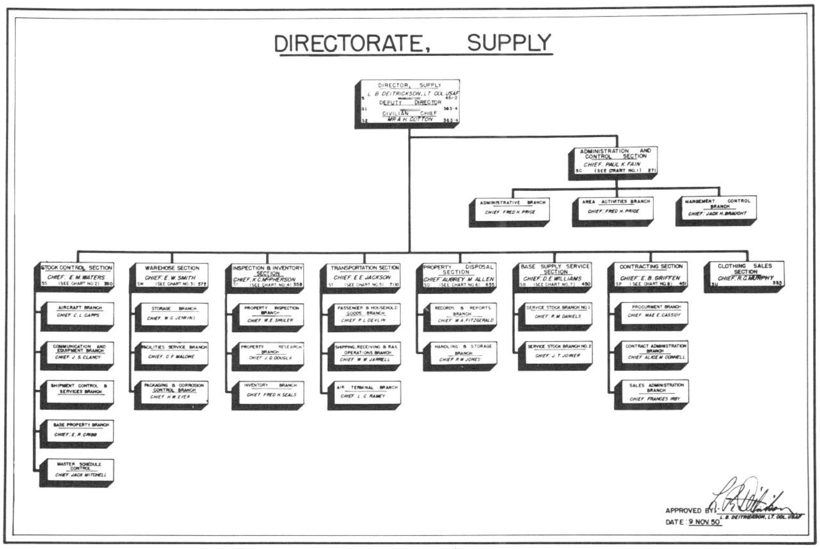

Many of those pictured above are shown in this organizational chart for the Directorate of Supply dated November 1950.

Maintenance Management

The 1950s also were years of change, new management techniques, and reorganization for the Directorate of Maintenance.

Originally named the Depot Engineering Department (according to the first Maintenance organization chart published December 17, 1942), Maintenance was redesignated the Maintenance Division on April 1, 1943. On June 21, 1948, it was redesignated the Directorate of Maintenance. It became the Directorate of Maintenance Engineering on March 1, 1952, and remained so throughout the 1950s.

From World War II until 1950 WRAMA had performed maintenance at three levels—organizational, field, and depot. Squadrons serviced their own aircraft and equipment on the flight line, inspected them before and after flight, and replaced parts.

Minor repair work was performed on jet engines in the maintenance shops at Robins in 1954.

Field maintenance performed in Robins AFB hangars and shops involved the minor repair and modification of equipment. WRAMA also completely overhauled aircraft and components and made major modifications.

In July 1951 the Air Force set up its first specialized maintenance depot at Kirtland AFB, New Mexico, to keep test equipment for atomic weapons ready for use. WRAMA's principal connection with the Atomic Energy program was through the Technical Services Division, Directorate of Maintenance Engineering, which handled reports pertaining to atomic energy. The division was responsible for suspension and release equipment for atomic weapons and, in this capacity, the Special Weapons Section was responsible for portions of the loading handbooks for all types of aircraft designed to carry atomic weapons.

In 1950 there was an immediate demand for aircraft, for additional wings, and for foreign aid. The B-29s which had been put into cocoons after World War II at Robins AFB were taken out of storage and reconditioned for service.

After 1950 the most important stress was placed on combat readiness. The Air Materiel Command organized about four thousand highly skilled civilians, including many WRAMA logisticians, into sixty mobile maintenance teams, ready to move at any hour to repair battle damaged aircraft or help Strategic Air Command bases get aircraft into action in the Korean War.

By late 1950, a shortage of experienced technicians caused HQ AMC to contract part of its maintenance workload to industry. This policy, as contrasted to the do-it-yourself philosophy of the late 1940s, was based in part on the experience of the industry-maintained Berlin airlift aircraft. It also evolved from a desire to keep the aircraft industry's muscles toned for possible wartime expansion.

At first, contract maintenance was restricted to second line aircraft and equipment and consisted of about ten percent of the depot-level maintenance workload. Contract maintenance grew to fifty-five percent of the total maintenance workload for the Air Materiel Command by 1958.

Meanwhile, the policy of making articles "like new" changed to one of "repair as necessary." WRAMA introduced IRAN (Inspect and Repair as Necessary) in 1953 to replace the DIR (Disassemble, Inspect, and Repair) concept which had begun in the early 1940s. Under IRAN, planes and equipment would be cycled through depots at set intervals to be repaired as necessary.

But IRAN, as first set up, did not prove as effective as anticipated and HQ AMC made major changes in 1958. In the future, the flow of airplanes through IRAN was to be determined by set agreements between AMC and the operating commands, with only stipulated work being done. Actual repair was to be the responsibility of the operating units, while WRAMA was to concentrate on the modification-modernization workload.

As the IRAN concept evolved, work increased at WRAMA. The Maintenance Propeller Shop expanded its overhaul facilities by taking over the shop space in the former vehicle repair building, Building 645. A serious problem with the propellers used on C-97 and KC-97 aircraft developed in 1956 and led to a program for retrofitting the entire fleet with nickel-plated and later dural blades. Involved were 1,036 spare blades at a cost of $41,615,204. More propellers were needed, but procurement had to wait for 1958 funds. The total cost of material for the retrofit was more than $63 million.

At WRAMA jet aircraft were inspected, winterized, safety checked, flight tested, etc., under Project High Flight, which began in March 1953. As of February 4, 1960, 3,282 High Flight aircraft had been processed at WRAMA. These included the RB-66 (B and C series), WB-66D, F-84, RF-84, T-33A, RT-33A, F-100, B-57, and F-86 aircraft.

The entire fleet of KC-97 aircraft (background) was retrofitted with nickel-platted propellers (foreground) in the maintenance shops in 1956.

RB-66 aircraft were prepared for flight overseas in the WRAMA maintenance shops during the 1950s. Standing in front of his aircraft is Captain James Gunnoe (now Lt. Col., retired).

As prime depot, WRAMA performed maintenance and modification work on the C-130. A prototype was produced and modifications completed by 1959. Programmed depot maintenance on the C-130 continues here today.

Earvin L. Joyner

The Deputy Director of the Directorate of Maintenance, a native Georgian, is one of the most highly decorated civilians in the United States Air Force.

Earvin L. "Pete" Joyner, a native of Brunswick, Georgia, received the AFLC Civilian Meritorious Service Award in 1970 and 1972, the United States Air Force Decoration for Exceptional Civilian Service in 1974, 1977, and 1979; and the city of Warner Robins. first "Good Government Award" in 1977.

Mr. Joyner is a veteran of the U. S. Army in World War II. He holds bachelor (1948) and master (1949) degrees in business administration from the University of Georgia and has attended the Industrial College of the Armed Forces.

Mr. Joyner came to Warner Robins, Georgia, in December 1956 as an operations research analyst in the Plans and Programs Office. He had been employed with the U. S. Bureau of Labor Statistics in Atlanta, Georgia, from October 1949 to January 1956, when he moved to Dallas, Texas, to serve as the bureau's southwest technical director from January to December 1956.

In February 1963, he was reassigned from chief of the Management Analysis Division to chief of the Plans and Programs Division. In December 1969 he was named Deputy Director of Plans and Programs. He became acting Deputy Director of Maintenance in September 1975 and Deputy Director in February 1976.

Mr. Joyner is a deacon in the Central Baptist Church, where he has taught an adult Sunday School class for twenty-five years, and is active in civic affairs in Warner Robins, where he lives with his wife, the former Sarah Hatfield of Monroe, Georgia. They have four children—Lee, Jeff, Bryan and Jennifer.

In order to supplement the operating commands' maintenance capabilities, the Air Materiel Command began a Mobile Maintenance Team program in 1954. Under this program WRAMA organized maintenance teams, and equipped and trained members for immediate emergency action.

The Air Materiel Command assigned maintenance workloads to Specialized Repair Activities (SRAs) which were single points for maintenance of particular functional subsystems. With advances in communications and automated data processing, these SRAs did not have to be physically close to the system managers, in the opinion of the headquarters staff. Efficiency increased when groups with similar maintenance requirements were assigned to one depot.

Among aircraft processed for overseas under Project High Flight was the T-33.

Management Assigned by Function Rather than Manufacturer

The 1950s featured shifts and adjustments in workload at WRAMA to seek to provide the most efficient overall logistical support to the Air Force possible.

One workload adjustment was the functional transfer of prime responsibility for the C-130 cargo transport aircraft from the Sacramento Air Material Area at McClellan Air Force Base in California to WRAMA on January 1, 1958.

The move was in line with the new HQ AMC policy announced in December 1957 of assigning, by family group, aircraft to the lead AMAs. Hitherto, one AMA had been prime for all of the aircraft produced by one manufacturer. Sacramento, for example, had been prime for all Lockheed aircraft—fighter as well as cargo.

To reduce time required to correct deficiencies in the C-130 and associated in-service equipment, WRAMA organized a Quick-Fix Team in June 1958. The team worked in close liaison with the Air Force Plant Representative Office and contractor personnel to improve the aircraft and correct service-revealed deficiencies.

As early as March 1958 C-130As were assigned to Pacific Air Force to replace C-119s. PACAF was scheduled to receive both thin and thick skinned configuration aircraft. Excessive noise and cracking of the skin on the aircraft was traced to the speed of the tip of the propeller blade, which raised the noise level above standards and created shock waves. These buffeted the fuselage and caused metal skin fatigue. Under the WRAMA modification program, thirty percent of the thin skin was replaced with a heavier material and the understructure was beefed up. Other modernization included changes to electrical, fuel, hydraulic, oxygen, fire extinguishing and air conditioning systems, flight controls, instruments and pressurization. The C-130 became the most versatile cargo aircraft in the Air Force.

WRAMA engaged in other major modifications to the C-130 during the late 1950s, including: equipping them with skis so they could be used in the Arctic and Antarctic areas; testing with a 116,000 pound payload to qualify as the largest assault transport in the Air Force; prototype installation of 450-gallon wing pylon fuel tanks to give the C-130 more range; and modifications to perform as a standard launcher/director platform for Q-2A, Q-2C and Q-4B drones in support of the Department of Defense's Weapon Systems Development Programs.

Prime responsibility for the C-130 was transferred from Sacramento Air Materiel Area to WRAMA on January 1, 1958. This is one of the earliest models of the C-130.

The C-130 Hercules aircraft were overhauled and modified in the large Robins AFB hangars by Directorate of Maintenance workers.

A C-130A undergoes modification in one of WRAMA's hangars.

C-130A aircraft are among many aircraft lined up on the Robins Ramp.

Maintenance workers perch atop the wing of an early model of the C-130A being repaired at Robins AFB.

A C-130B turbo-prop engine is repaired by maintenance employees at Robins.

A C-130A is equipped with skis for use in the Arctic and Antarctic areas. This modification was performed by WRAMA employees.

WRAMA maintenance workers modified C-130B aircraft for jet-assisted takeoffs.

C-130s were modified by WRAMA employees who added satellite tracking equipment in the aircraft.

Procurement Management

In the 1950s Americans were enjoying an unprecedented living standard. Purchasing power had doubled since the 1920s.

But with the exception of the Korean War years, military budgets of the 1950s were severely constrained. Defense budget cuts for the fiscal year ending June 30, 1953, before the Korean truce was secured, totaled $4.3 billion, the largest dollar cut in any defense budget for any year since 1946.

Thus, WRAMA had to produce more with less as it entered the decade of the 1950s. It was during these years of challenge that Procurement became a major directorate.

Since early in 1942, the Purchasing and Contracting Section of the Supply Division handled local purchasing and contracting for Air Corps supplies, equipment, and services. In middle 1951, the purchasing and contracting functions of supply were transferred to a newly established Procurement Office.

In early 1952 the Procurement Office was made a directorate, an indication of the increasing importance of WRAMA's procurement responsibilities. Procurement's organizational structure expanded in July 1952 and again in November of that year when existing functions were regrouped and a detailed structure to accommodate prime class buying was established.

In parallel with procurement decentralization to the AMAs, a new system of appropriation accounting was begun July 1, 1952 which involved the use of International Business Machine's accounting equipment for processing records and reports for fundings.

Two memorable events occurred in 1953. The first was a near-tragedy. On January 14, 1953, the procurement building burned and almost all records were destroyed.

But, despite loss of the building, procurement employees earned for WRAMA the Nelson S. Talbott Award for outstanding work in the procurement of anti-friction bearings.

The next year, on July 1, 1954, procurement responsibilities for the states of North Carolina and Virginia were transferred from Middletown AMA at Olmsted AFB, Pennsylvania, to WRAMA.

From a small twenty person office in 1942, procurement has grown to a directorate (now called the Directorate of Contracting and Manufacturing) of seven divisions, each with several branches. In 1942 there was an annual workload of six thousand contracts. Today there are 550 employees handling 70,000 contract actions.

Mission support has grown from furnishing purchase information, orders, and directives to subdepots, to providing contract support on a worldwide basis.

Pricing and contract buyers handle Procurement business in their office in Building 215. The Procurement Office became a directorate in early 1952.

Materiel Management

Following the successful conclusion of the decentralization program in 1953, by a simple evolutionary process, there began to be overtones of systems management at WRAMA. This was a new term, a new kind of operating philosophy generated partly because there was no remaining central focal point to bring all these functions together such as HQ AMC at Wright-Patterson AFB, Ohio.

The expression WSPO (Weapons System Program Office) was first heard with the emergence of an organizational component at Robins in the old Plans and Programs Office. Here, weapon system managers were identified to a given weapon system such as the B-57 or TM-61, or some other system.

These WSPOs brought together, principally in terms of statistical data, the support posture of individual weapon sytems. The operation was not ideal, however, because of the reactive nature of such an arrangement. System health required surveillance on a "real time" basis.

In September 1958, HQ AMC directed WRAMA and the other AMAs to create a Directorate of Logistics Support Management by January 1, 1959.

Technical Services Management

Among the Maintenance functions transferred from HQ AMC to the Air Materiel Areas following decentralization was the function of technical services management.

Prior to decentralization, WRAMA had only the distribution and repair responsibilities for certain items. WRAMA did not have the technical support responsibility. Now WRAMA was able to determine how often items might wear out and determine when or how much to buy. Instead of being merely a point of distribution—an inventory control point—WRAMA became manager of an item.

Furthermore, before decentralization, there was fragmented responsibility for item and system management—part was at HQ AMC and part at WRAMA. This ended when technical services management came to WRAMA.

In the production management area, the action resulted in better programming, better shops scheduling, and better shops flow control involving everything for which WRAMA had repair responsibility.

The Technical Services Division remained in the Maintenance organization at WRAMA until the Directorate of Logistics Support Management was established.

This Maintenance Directorate technical services function later became the nucleus of the engineering, reliability and maintainability, and production management functions, which we find today in the Directorate of Materiel Management.

Technical Services employees in the Directorate of Materiel Management maintain close support of assigned electronics systems.

The WRAMA logistics management operation featured separate divisions for major weapon systems, for example, the B-57, TM-61, or the TM-76 Divisions, all of which heretofore had been in the Directorate of Supply and Services with parallel functions in the other operating directorates.

This organizational arrangement was adopted throughout the AMC on January 1, 1959, as the structure established at WRAMA became the standard pattern for the command. From Supply they transferred responsibilities for requirements and distribution and for cataloging and standardization. Technical services and production management came from Maintenance.

Major weapon systems assigned to WRAMA at the time included the C-130, B-57, C-118 aircraft, the TM-61 Matador and the TM-76 Mace missiles. Accessories included propellers, weapons, and bearings; and armament included bombing-navigation and fire control systems. Cataloging, Plans and Programs, and the Customer Support Office were supporting functions.

The Directorate of Materiel Management came into being on January 1, 1959 (though it was called the Directorate of Logistics Support Management until March 1, 1960). Materiel Management became the hub for WRAMA's worldwide logistic support to assigned weapon systems, support systems, and prime commodities. Local industrial operations remained under the Directorates of Maintenance Engineering and Supply and Services.

The directorate's primary organizational objective was to consolidate under one director the activities involved in making materiel requirements decisions. A corollary objective was to separate the new director from the taxing but different job of running a large physical plant.

Among the responsibilities centralized in the new organization were those for directing the repair program, the financial portion of requirements, and in service engineering for assigned systems and items.

Overseas Depots

Going into the 1950s, the Air Force had overseas depots. We had had them during World War II and some remained at Burtonwood, England; Chateauroux, France; Nouasseur, Morocco; the Philippines, and Tachikawa, Japan.

Throughout the entire history of the Air Force, our forces in the field were better supported at this time than ever, even in times of war. But it was expensive—in terms of materiel aggregation and the amount of personnel required.

The Air Force decided to close the overseas depots and to resupply its operating bases throughout the world directly from its continental United States depots—the first branch of the service to do so.

Late in 1959 the WRAMA Commander, Maj. Gen. A. V. P. Anderson, Jr. and the deputy director of the Directorate of Logistic Support Management, A. H. "Hal" Cotton, made an around-the-world logistics survey for the Air Force. It would determine the degree of logistics support provided our forces in the field.

The report WRAMA rendered and the accolades given the entire air logistics systems were unparalleled in the annals of logistics support. The WRAMA team found a high degree of activity regarding the amount of attention given to weapon systems and the amount of support provided to them.

However, with the beginning of the decline of the U. S. military budget and in the advent of the days of Robert McNamara as Secretary of Defense, these functions began to phase down in terms of materiel aggregation and in terms of individual support provided to our operating forces.

It did not in any way, however, diminish the functional responsibility that the AMAs had in providing this support to our fighting forces wherever they might be located.

Training programs covering a wide range of topics were conducted regularly for employees of the Directorate of Materiel Management.

Service Engineering Division draftsmen produced electronic schematic diagrams, line drawings, and design modifications.

The Pack Rat of Robins

It was called the Pack Rat squadron, characterized by a drawing of Mighty Mouse carrying a log. This was the symbol of one of the busiest and most highly respected tenant organizations in the history of Robins—the 7th Logistics Support Squadron.

It was a fitting symbol since members of this squadron, assigned to WRAMA on October 18, 1954, were constantly on the move packing materiel and carrying it to far away places aboard C-124 Globemasters.

Thousands of Pack Rat symbols blanketed the base. The 7th Loggers put stickers on every available barracks and transient aircraft they found while on temporary duty in the states and overseas. It eventually became an international symbol—the 7th Loggers' answer to "Kilroy was Here" of World War II fame.

It was a gung ho outfit. Though its members had to fly for more than nine days with just three days rest, aloft all day and loading and unloading materiel at night, they had pride in their C-124s, known affectionately as "Old Shakeys." They had to fly at altitudes of some five thousand feet because the aircraft were not pressurized. The Globemaster looked like a bumble bee with a fat body and short wings—a plane which many had predicted would never fly.

The 7th changed its name to the 58th Military Airlift Squadron in early 1966 and converted to the C-141 beginning in 1967. The 7th Loggers, while known as the 58th Military Airlift Squadron, completed three hundred thousand accident-free flying hours spanning fourteen years of operation (1957-1971) since the tragic accident of March 2, 1956. The 7th LSS lost one C-124 when it crashed upon return to Keflavik, Iceland, killing all seventeen crew members aboard.

Just months before the squadron's deactivation, it was selected as the Outstanding Airlift Squadron in the Military Airlift Command for 1970. Then on August 15, 1971, a seventeen-year-old courtship with Robins AFB came to an end when the 58th MAS was deactivated. Its members were absorbed into MAC units throughout the Air Force.

A crew from the 7th Logistics Support Squadron stands in front of a C-124.

1831st GEEIA Squadron

Members of an organization which was a member of the Robins team for almost two decades really never spent much time at their host base.

These 1831st Ground Electronics and Engineering Installation Agency (GEEIA) Squadron personnel were too busy on temporary duty in far-flung points such as the Azores, Belgium, and Puerto Rico to become well known at Robins. They installed and maintained facilities and systems of the Air Force ground communications electronics meteorological program.

The squadron known as the 2860th GEEIA Squadron, was originally formed at Robins AFB on January 1, 1959, composed of personnel from the 4th Communications Construction Squadron; Detachment 3, 1889th Army Airways Communications System (AACS); and electronic and communication workers in the Directorate of Maintenance at Robins Air Force Base. It became a part of the Air Force Logistics Command. Previously it was known as the 1889th AACS Squadron and was located at St. Louis, Missouri.

This well traveled squadron remained an AFLC unit until April 1, 1970, when it was transferred back to the Air Force Communications Service. It became a part of the Southern Communications Region under AFCS rather than the Eastern GEEIA Region under AFLC. Its name was changed from the 2860th GEEIA Squadron to the 1831st GEEIA Squadron, which it kept through the remainder of its stay at Robins.

On June 30, 1977, the 1831st GEEIA Squadron was absorbed into the 1926th Communications Installation Group, which remains a tenant organization at Robins today.

The 2860th Ground Electronics and Engineering Installation Agency was located in Building 994. It previously was the 1831st GEEIA Squadron.

The Reserves—A Proud Heritage

The only major command ever headquartered at Robins Air Force Base, the Air Force Reserve, boasts a proud heritage.

When the forerunner to AFRES, the Fourteenth Air Force, one of four Continental Air Command numbered air forces, transferred to Robins from Orlando, Florida, October 28, 1949, it brought with it a brilliant record compiled while fighting the Japanese as the "Flying Tigers" in the China-Burma-India theater in World War II.

Formed in the summer of 1941 by a retired Air Corps officer with long experience in China, Claire L. Chennault, it was originally known as the American Volunteer Group. Flying ancient, patched-up P-40s, the "Flying Tigers" piled up an astonishing total of 297 confirmed air victories between December 18, 1941 and July 4, 1942. They took a Japanese toll conservatively estimated at fifteen hundred pilots, navigators, gunners, and bombardiers.

The insignia for the Fourteenth Air Force, then Continental Air Command at Robins.

The volunteer group became the China Air Task Force on July 4, 1942 and continued its outstanding record through improvisation, ingenuity, and dogged determination.

After the war, the Fourteenth Air Force became a part of the Continental Air Command (CONAC) which administered and trained Air Reserve Forces units and individuals. The Fourteenth Air Force moved to Robins Air Force Base in October 1949.

The Fourteenth Air Force moved into its new headquarters, Building 210, in early 1953.

Located in several buildings scattered over Robins, the Fourteenth didn't move into its new headquarters building, Building 210, until early in 1953. In the latter half of 1950 and almost through 1951 the Fourteenth became involved in the Korean War as CONAC called numerous reserve and guard units to extended service.

In January 1957, amid rumors of moving to another installation, and as the entire Defense Department began to feel the Eisenhower economies, the Fourteenth Air Force reduced and consolidated its holdings on Robins Air Force Base. The famed "Flying Tiger" Band was discontinued. The Fourteenth Air Force now had supervision over reserve and guard activities in twenty-two states stretching from Canada to the Gulf of Mexico.

The insignia for the Continental Air Command which moved its headquarters to Robins in April 1961.

On September 1, 1960, the Fourteenth Air Force was inactivated. CONAC discontinued the three numbered air forces and replaced them with six Air Force Reserve regions. At Robins, the 3rd Air Force Reserve Region replaced the Fourteenth Air Force. But just four months later, on the weekend of March 11-12, 1961, the 3rd Region moved to Dobbins AFB, Georgia. Headquarters CONAC then made its move to Robins from Mitchel AFB, New York, in middle April 1961. Considering economic and political factors, plus encroachment, which was strangling the installation, the Air Force had decided to close Mitchel.

While the 1950s was the decade of the Fourteenth Air Force at Robins AFB, the 1960s was the decade of the Continental Air Command.

It was as a mobilization force that the post-Korean Air Force Reserve had been intended. With the stiffening of Cold War attitudes as a result of new tensions in Berlin in 1961, five squadrons of Air Force Reserve C-124 troop carrier units were ordered to active duty on October 1, 1961, and released on August 27, 1962.

Two months later, eight C-123 and C-119 equipped troop carrier wings and six aerial port squadrons were recalled to active duty as a result of the discovery of Russian missiles in Cuba. The units, all of which served at home stations, were released on November 28, 1962.

Early in 1965 the reserves also contributed heavily to the airlift necessitated by U. S. intervention in the Dominican Republic.

Inevitably, the situation in Southeast Asia necessitated the mobilization of some Air Force Reserve units in January and May of 1968. In January, two military airlift wings and five groups equipped with C-124s and an aerospace rescue and recovery squadron flying HC-97s were ordered to active duty. On May 13 a tactical airlift group minus some support elements, three aerial port squadrons, a medical service squadron, and an aeromedical evacuation squadron were recalled. All had been released by June 1969.

Continental Air Command went out of existence when Headquarters Air Force Reserve was activated at Robins on August 1, 1968. The new agency was a result of a comprehensive reorganization of the Air Force Reserve management structure. It became a special staff agency responsible for directing the Air Force Reserve field structure and in concert with the Air Staff for devising policies, programs and plans for the Air Force Reserve units.

The Air Force Reserve replaced the Continental Air Command at Robins in August 1968. This is their insignia.

While the primary role of AFRES was to prepare Air Force Reserve flying and nonflying units to augment the active force when needed in peace or war, it also inherited other responsibilities from the predecessor command, CONAC, most of which involved coordination and cooperation with other branches of the military and other governmental agencies.

Along with the headquarters, four other AFRES organizations are located on Robins. These include the 581st USAF Band which accompanied CONAC to the base in April 1961; the 94th Aerial Port Squadron, activated on July 1, 1975; the 402d Combat Logistics Support Squadron (CLSS), activated October 1, 1978; and the 402d Civil Engineering Squadron, activated January 1, 1982.

The 94th Aerial Port Squadron trains in peacetime to achieve the wartime capability to provide cargo, passenger, mail, and patient handling services for all airlift aircraft serviced wherever the unit is assigned when it is mobilized.

Master Sergeant Jamey Loy of the 402nd Combat Logistics Support Squadron, an AFRES unit at Robins, checks repairs to an AFRES F-105 aircraft.

Clockwise, these Air Force Reserve aircraft are: A-37, F-4, A-10, and F-105.

The 402d CLSS provides worldwide military teams which can be deployed to perform aircraft battle damage repair, combat packaging, and supply operations. It also provides teams to accomplish depot level technical assistance and field support to include modification, repair, and maintenance of aircraft.

The 402d Civil Engineering squadron trains in peacetime to assure its Prime BEEF (Base Engineer Emergency Forces) teams can meet the wartime base engineering requirements of the Air Force Logistics Command.

An AFRES Security Policeman guards an "intruder" during a simulated terrorist attack in an exercise.

An Air Force Reserve C-119 "Flying Boxcar" takes off from a CONAC installation in the Southwestern United States.

Air Force Audit Agency Office

The Air Force opened a Resident Auditor's Office at Robins Air Force Base in 1950. It was known as the Auditor General and came directly under Headquarters Command, U. S. Air Force.

Today, it is a separate operating agency called the Air Force Audit Agency (AFAA) with headquarters at Norton Air Force Base, California.

Its present office on the third floor of Building 125 has grown from a five-person staff to its present size of thirty-one persons and its audit coverage has grown considerably. In early years it was limited to property and sales audits which included supply, disposal, and commissary activities. In the middle 1960s, centrally directed audit programs were developed, placing emphasis on weapon systems and those areas where most of the Air Force money was spent. This approach is continuing today. The Robins office won the Outstanding Office Award in 1977 and in 1979 for the Air Force Audit Agency.

Detachment 5, Headquarters Air Force Communications Command

Detachment 5, Headquarters Air Force Communications Command (AFCC), located within Building 210, serves as the Communications-Electronics Support Office for Headquarters, Air Force Reserve (AFRES). First known as the Directorate of Telecommunications within DCS/Operations, Continental Air Command (CONAC), the functions of the office moved to Robins from Mitchel AFB, New York in the Spring of 1961.

In July 1962, the CONAC Directorate of Telecommunications was discontinued and the Office of CONAC Command Communications (Detachment 2, Headquarters Air Force Communications Service [AFCS] was established. The new office was to be responsible for providing communications-electronics staff support for HQ CONAC on matters pertaining to command and base communications, flight facilities, and air traffic control facilities which AFCS would operate for CONAC.

In August 1968, CONAC became AFRES, but Det. 2, AFCS kept its name. In March 1974, Det. 2, AFCS was discontinued and Operating Location (OL)-B, AFCS was established with no changes in its responsibilities. On August 1, 1977, OL-B, AFCS deactivated and became Det. 5, AFCS the following day.

The last change occurred on November 15, 1979 when the detachment participated in the redesignation of AFCS to AFCC. The unit became Detachment 5, HQ AFCC. The change was announced by General Lew Allen, Jr., Air Force Chief of Staff.

Detachment 5 is also the AFRES/DC—or, the Directorate of Communications for HQ AFRES. There are four functional divisions within AFRES/DC: The Telecommunications Division watches over all communications operations and maintenance matters; and is AFRES' radio frequency manager. The Reserve Communications Forces Division plans, approves and coordinates the activities of approximately 1,033 reserve communications personnel at 34 geographically separate locations. The Plans and Programs Division initiates, coordinates, consolidates, and approves all communications-electronics planning by and in support of reserve units. This includes operational as well as equipment installation on modification plans. The Air Traffic Control Division oversees all ATC activity at AFRES bases, and by or for AFRES units. It initiates changes in policy or procedure to ensure safety of AFRES operations along with mission accomplishment.

3503d U.S. Air Force Recruiting Group

The 3503d United States Air Force Recruiting Group began as the 3503d Processing Center at Moody Air Force Base, Georgia, on April 15, 1952. The processing center was renamed the 3503d USAF Recruiting Group in July 1954 and moved to Robins Air Force Base, Georgia, where it remains today.

The group emphasizes community involvement and works with charitable organizations, civic groups, and national organizations, such as the Boy Scouts.

The group was named the top recruiting group in the nation in 1978, 1979, and 1980.

Base Newspaper

On March 21, 1942, a single sheet newspaper called the *Robins Call* made its first appearance at Robins Field. Ten days later, the size of the publication jumped to five pages.

Among interesting items to be found in the early editions were stories about how enlisted men, regardless of rank, could get married without the consent of their commanding officer; a description of what was then the new Army Air Forces shoulder patch; and about how off-duty salutes, which up to that time were optional, were now mandatory.

On July 31, 1942, the *Wellston Beacon* was first published by the Civilian Welfare Association of the field for the benefit of civilians.

The *Robins Call* came out with a 12-page multilith edition, an official publication for military, on January 15, 1943. It carried the nationally-syndicated *Male Call* strip, of interest to the military. On February 12, 1943, the *Robins Call* was replaced by the *Robins Field News.*

On April 7, 1944, the *News* combined with the *Beacon* to become the *Robins Field News-Beacon,* published for both military and civilians.

On June 8, 1945, the *News-Beacon* became the *Robins Field News* with the announcement that the paper would devote itself to matters of interest to GIs only. All civilian coverage was dropped. It was renamed the *Robinews* on June 29, 1945. On March 1, 1946, the *Robinews* came to an end and the base newspaper was no more.

After the World War II years, on December 1, 1947, the first edition of the *Robins Times* appeared. Sponsored by the Civilian Welfare Council, it was an unofficial publication, since it contained paid advertisments and was published for civilians. The *Times* became an official Air Force publication (no advertising) on January 29, 1952, and was published for military and civilians. It was then published by the Information and Education Unit at Robins. Publication ended on June 23, 1953.

Following the demise of the *Robins Times* a new publication—the *Robins Rev-Up*—made its appearance on October 1, 1954. Cost was borne by the Military Non-Appropriated Central Post Funds and the Civilian Welfare Council. No advertising was permitted and the Information Services Office was responsible for its content.

On August 15, 1957, it became a commercial enterprise newspaper, which meant that its cost was paid by advertisements. The *Rev-Up* featured numerous personality columns, such as "Hawkeye," by Ed Hawkins,

…OUT WELFARE MEMBERSHIP CAMPAIGN OPENS

HE WELLSTON BEACON

WELLSTON, GEORGIA, FRIDAY, JULY 31, 1942 No. 1

…bins Field Employees Asked To Rent Spare Rooms

…NDATION

…jor Altmayer …reat Asset …o Robins Field

Major M. S. Altmayer, the amiable gentleman who serves as Public Relations Officer for Robins Field and Wellston Air Depot, deserves commendation from civilian and military personnel at the post for the many favors that he has done for their good.

Yes, civilian and military personnel have many things to thank Major Atlmayer for.

When Colonel C. E. Thomas, Jr., commanding officer of the posts, has something that he wants done he calls on Major Atlmayer, and the PRO sticks with the job until it has been completed.

The popular major enabled the personnel to get their gas rationing cards at the post so that they would not have to rush to Macon or their home town to get the cards issued after working hours; he went to Washington, D. C. and helped get 250 houses for Wellston and right now he is busy trying to get 250 more houses, and if the Major's ways run true to form he will get the additional houses; he has done much toward getting a new road and the present road leading to Wellston improved; and he has done much to get bus and railway bottlenecks remedied to and from Wellston.

Major Altmayer is the gentleman who arranged to get the military access highway signs up between Macon and Wellston. He had the speed limit placed at 40 miles per hour and he got the city council in Macon to allow no parking on Houston Avenue between Wellston and where Broadway meets Second Street.

The above are just a few of the many great things Major Altmayer has done for civilian and military personnel at Wellston. He enjoys doing things. . . . If he can be of help to a person he will do it or die.

HUNDRED PERCENT IS SET AS GOAL BY LEADER MILES

Best Welfare Association In World Is Aim of President McPherson

An all out drive for memberships in the Robins Field Welfare Association will be launched Monday, according to W. C. Miles, chairman of the membership committee.

"We have done well in our drive thus far," Mr. Miles said today, "but we can do better . . . and we are going to. We will strive to get every civilian employee at Robins Field, Wellston Air Depot to join the association."

General Chairman Miles has appointed a person as a membership chairman in each department. His department appointees have been asked to form committees to help stimulate the sales of membership

(See HUNDRED, Page 15)

Vacancy Filled By Geo. Kennedy

George Kennedy, senior instructor in vocational training section at Wellston Air Depot, was appointed a member of the Robins Field Welfare Association's Executive board today.

Mr. Kennedy replaces Captain A. Mack Dodd who resigned recently when commissioned in the Army of the United States. Mr. Kennedy is chairman of the association's resolutions committee.

The new board member attended the Mercer University School of Law in Macon and later practiced law in Augusta with his brother Fred Kennedy.

In 1937 Mr. Kennedy was appointed assistant director of the State Welfare Department. While on that job Mr. Kennedy and his family made their home in Atlanta.

COLONEL CHARLES E. THOMAS, JR., commanding officer of Robins Field and Wellston Air Depot, is honorary president of the Robins Field Welfare Association. When the association was organized Colonel Thomas was presented with membership card No. 1. The welfare association dedicates this first issue of The Wellston Beacon to the commanding officers. (Macon Evening News Photo.)

EDITORIAL

Your Newspaper

Today the Robins Field Welfare Association takes another step forward in presenting this newspaper, your newspaper . . . The Wellston Beacon.

The Wellston Beacon is a civilian enterprise published in the interests of the citizens of Wellston and the personnel of Robins Field. This is your newspaper. It is your spokesman. Every employee at Robins Field and each citizen of Wellston has a share of stock in this publication.

The Wellston Beacon will be issued on the first and third Friday of each month to every person, military or otherwise at Robins Field and to other citizens of Wellston.

Policies and statements reflected in the news columns or in editorials will be the views of the editors or other writers and under no circumstances are to be considered as representing official views of the commanding officer of Robins Field, Wellston Air Depot, the United States Army Air Forces or any other branch of the government.

All articles appearing in The Wellston Beacon are prepared by members of the Robins Field Welfare Association outside of their working hours. None of the work for this paper is prepared on government time. The writers, all interested in helping the morale and welfare of their fellow workers, contribute their time and writ-

(See EDITORIAL, Page 15)

HOUSE KEEPING UNITS SOUGHT FOR NEWCOMERS

Employees of Robins Field and Wellston Air Depot who have spare rooms in their homes in Macon and other nearby towns were urged today by Sam F. Mayer, welfare and housing director for Robins Field, to assist with the housing problems which face officials and workers at the large war plant here.

"There are many ways in which our employees and other citizens of Middle Georgia can help us with our housing headaches", Mr. Mayer said. "There probably are hundreds of Robins Field workers who have spare rooms which they could rent to personnel who are in need of a place for light house keeping.

"If a worker does not have a vacancy in his home he or she could persuade a neighbor or some other good friend to rent their spare room to and share their kitchen with Wellston workers. The barracks and the 250 houses which will be built here will not come near taking care of our housing needs".

Rooms Will Have Job

In many towns where there are war plants people who never had a stranger in their homes have fixed up rooms where their children once slept to furnish housing for war workers.

Spare rooms in homes will have a war job to do.

A recent report from Washington states that housing must be secured in private homes for not less than 850,000 war workers be-

(See HOUSE, Page 15)

GROUND BROKEN FOR 250 HOUSES IN ROBINS MANOR

The initial major step toward the construction of a city of homes and assembly places here was taken Tuesday when ground was broken and surveyors began laying off the large plot where 250 homes and a recreation area will be bulit.

Major M. S. Altmayer, Robins Field public relations officer, announced the actual beginning of work on the model city here.

Last week the Wellston Housing Corporation began building a warehouse and cutting plant on the right-of-way here.

The homes will be constructed on the property of Homer J. Walker.

The farthest point in the village will be only six-tenths of a mile from the gates of Robins Field. This will contribute to much relief on the gasoline and tire shortages.

Office Quarters

The Homer J. Walker home is being converted to office quarters for the construction company

Major Altmayer says a recreation area has been set aside in the village's area and a playground, baseball park, tennis courts and many other amusement spots will be made. Efforts are now under-

(See GROUND, Page 15)

EXECUTIVE BOARD SLATED TO MEET FRIDAY, AUGUST 7

The executive board of the Robins Field Welfare Association will hold its August meeting at 12 o'clock noon next Friday.

The meeting will be the first for the board since its organization session, June 23.

Members of the board include Karl McPherson, chairman; George Alexander, Malcolm G. Bell, Wallace Durrett, Miss Rosita Gostin, Mrs. Christine Lidberg, W. C. Miles, Stockton Odom, Miss Juanita Rice, Wilbur Roach, Arthur Schroeder, and Miss Pearl Wilson.

All members of the board were appointed temporarily. They will hold their places on the board until the general election next January.

AN EXPERT'S OPINION—

Dispatches Indicate Extensive Employment Of Field Artillery By Germans

By Major George Fielding Eliot
(Copyright, 1942, N. Y. Tribune, Inc.)

The dispatches from the Russian front indicate a very extensive offensive employment of field artillery by the Germans. Again and again occurs the phrase "German artillery and aircraft attacked." This is significant, if we stop to think about it, for the speed of the German advance in the bend of the Don has been such that the artillery is certainly not attacking from fixed positions, nor is it being displaced forward by a succession of bounds from position in the traditional manner of artillery supporting an advance. It is being used offensively—that is to say, guns on self-propelled mounts are being pushed forward with the leading elements of motorized and mechanized troops, adding their volume of fire to that of tank guns and of aircraft, particularly dive bombers. This heavy volume of close-supporting fires probably accounts for the failure of the Russian rear guards appreciably to delay the German advance in some sectors. There is however nothing to support the further supposition that artillery has been used offensively against Russian tanks; it may or may not have been.

Observers Inclined

Some observers are inclined to believe this German use of artillery to be a development of the tactics successfully used by Marshal Rommel in Africa, where he sent artillery on self-propelled mounts forward with his tanks in close support, and indeed in his latest operations made no large tank attack without strong accompanying artillery support. This use of his artillery, particularly the 88-mm. All-purpose gun, was an important feature of his recent successes. It was particularly effective when closely coordinated with squadrons of dive-bombers.

This offensive use of field artillery is, however, not a German invention. We have become so accustomed to think of the Germans as originating tactical ideas, and of other people as merely countering each new German method, rather belatedly, with a defense against

(See OPINION, Page 12)

Volume I, Number 1, *of the* Wellston Beacon, *July 31, 1942.*

The Robins Rev-Up *first appeared October 1, 1954.*

a supervisor in the Directorate of Materiel Management. The "Hawkeye" column appeared in the late 1950s.

An example of the humor of these early columns is shown in this excerpt from a "Hawkeye" column dated September 12, 1957. In recounting a greyhound race he watched in Florida, Hawkins said:

. . . Them dogs didn't look like any hounds I ever saw, and they wasn't gray either. In fact—they must have been beginners 'cause they had numbers on the dogs. Folks 'round here could tell their dogs by the tones of their barks.

Then them nearsighted dogs took off after a stuffed rabbit on a track. . . .

I also found out you could make a little wager if you wanted to—(them folks sure was took, 'cause them dogs never did catch that rabbit).

When I get back home I'm gonna rent the race track, put a flower garden in it, get me a real rabbit, and make myself a fortune—

The Robins Rev-Up continues to be published as an unofficial base newspaper for both military and civilians at Robins Air Force Base.

Some Talented People

Three nationally famous men—a painter, Butler Brown of Hawkinsville, Georgia; a cartoonist, Johnny Hart of Nineveh, New York; and a photographer, Chuck Mantelli of Rochester, New York—all worked at Robins in their earlier years.

Butler Brown, who was a computer programmer at Robins until his retirement in May 1971, had two of his paintings hanging in the White House during the presidency of Jimmy Carter. Butler painted the scene that is depicted on the jacket of this history.

Hart was stationed at Robins in the early 1950s as a photographer. He is the creator of the nationally-acclaimed cartoon strips, *B. C.* and *Wizard of Id*, which have won national and international awards. Some of his comic strips, both *B. C.* and those he did while at Robins, are used in this history.

Mantelli, coincidentally a corporal like Hart who served with him in the WRAMA Public Information Office in the early 1950s, has won numerous awards in the field of photography. Several of his photographs, taken while he was stationed at Robins, are used in this history.

Talent show of 1951 at the Service Club featured two young airmen, Johnny Hart on the left and Chuck Mantelli on the right.

Map of Robins Air Force Base

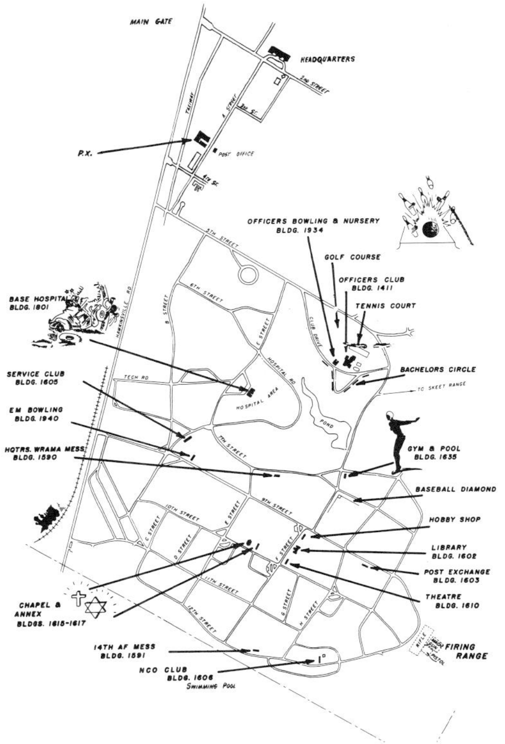

Among the projects on which Hart and Mantelli worked was a guide for new military personnel in 1951. This is the map which was featured in that work.

Upon graduation from high school in 1956, Butler Brown began work at Robins. He began painting in 1968 and after his retirement in 1971 devoted all of his time to painting scenic pictures of old farm buildings and rural landscapes in middle Georgia.

His works received statewide attention when then Governor Jimmy Carter saw his paintings at two exhibitions in Macon, Georgia. Governor Carter told Brown that he liked his paintings very much "because he could relate to them," Brown recalls. Brown lent him a painting to hang in his office in the state capitol.

Butler Brown gained national prominence when Rosalynn Carter presented two of his paintings to her husband for Christmas in 1976 and they hung in the White House throughout the president's term. The King of Saudi Arabia, the Minister of Culture in the People's Republic of China as well as Ambassadors and United States Senators also have included his paintings in their collections.

He has held over fifteen one-man exhibitions from Florida to Virginia and has had original lithographs published by The Franklin Mint, Circle Galleries, Ltd of New York, American Express Company and the James Shuptrie Galleries of Signal Mountain, Tennessee. He has won numerous awards and honors and was listed in *Who's Who in America* and *Who's Who in the South* in the 1978-1979 and 1980-1981 editions.

Butler and his wife Laverne, who is employed at Robins, live in Hawkinsville, Georgia. They have two children.

Butler Brown and his wife Laverne pose during a reception for General and Mrs. Bryce Poe II, during which Brown's paintings were displayed.

Butler Brown gained national attention when Mrs. Carter selected a Butler Brown work to give to the President-Elect in 1976. Mr. Brown and his wife Laverne are greeted in the White House by the President.

A graduate of Ediston Tech High School in Rochester, New York, where he studied photography, Chuck Mantelli worked as a commerical photographer until 1950 when he joined the U. S. Air Force. He was assigned to Robins between 1950 and 1953, where he served as base photographer, specializing in area photos.

After his discharge from the Air Force in October 1954, Mantelli continued his education in photography at the University of Rochester and at the Winona (Indiana) School of Photography.

He then started his own business, specializing in advertising-communication, fashion and food. In 1964, he won the awards for best color and best print in the eleventh annual International Day Competition at Niagara Falls, New York, for the *Courier Journal* in Rochester. Mantelli has also won numerous other awards in photography. He is a member of the Rochester Advertising Council and the Professional Photographers of America.

Mr. Mantelli and his wife Jerri live in Rochester, New York.

Chuck Mantelli, Robins Airman in 1951, gained acclaim as a prize-winning photographer in upper New York State. Several of his photographs taken in the early 1950s are used in this book.

A native of Endicott, New York, Johnny Hart was assigned to Robins in the early 1950s. Here he met and married his wife, Bobby. Fittingly, pictures of their wedding in the base chapel were taken by his buddy, Mantelli.

After his discharge in 1954, Hart sold his first cartoon to the *Saturday Evening Post*, heralding the beginning of his career. But he needed steady work so he took a job with the art department of General Electric Corporation.

In March 1958, Hart began the cartoon, *B. C.* Two years later he created the *The Wizard of Id*, which lay dormant for several years until it appeared in 1964 with Brant Parker as its artist. Hart credits Parker with being a prime influence in his making cartooning a profession.

Hart's cartoons won numerous awards—best humor strip (*B. C.*) in 1967, cartoonist of the year for 1968, best humor strip in 1971 (*Wizard of Id*), best animation film (*B. C.*) in 1973, all of which were presented by the National Cartoonist Society.

He also was named the best cartoonist by the International Congress of Comics, Lucca, Italy, in 1970, making him the first American cartoonist to win the award.

He won France's highest award for cartoonist of the year in 1971; a public service award in 1972 from the National Aeronautics and Space Administration; the Silver Bell Award for best television commercial-animation (*B. C.*) in 1973 from the National Advertising Council; and the Golden Spike Award for best animated TV commercial (*B. C.*) from the international Society of Radio and TV Broadcasters.

Mr. Hart and his wife Bobby live in Nineveh, New York. They have two children.

Johnny Hart, Robins Airman in 1951, gained national acclaim as creator of the comic strips B. C., *which he continues draw and the* Wizard of Id, *drawn by his friend Brant Parker. Johnny provided the cartoon above especially for this Robins AFB Pictorial History.*

A PHOTOGALLERY: *Leaders of the 1950s*

Commanders, Warner Robins Air Materiel Area[3]

Kingston E. Tibbetts was born in Columbus, Ohio, February 17, 1907. He was educated in the public schools of Columbus and attended Ohio State University.

General Tibbetts entered military service as a flying cadet at Fort Hayes, Ohio, in October 1927, and was commissioned a second lieutenant in the Air Reserve. He returned to Columbus, where he worked for the Curtis Flying Service until he received his regular commission, May 2, 1929, and was assigned with the First Pursuit Group, Selfridge Field, Michigan.

During the 1930s, Tibbetts was a flight instructor at Brooks Field and Randolph Field, Texas. He also served with the 18th Pursuit Group at Wheeler Field, Hawaii, and the 9th and 22nd Bomb Groups at Mitchel Field, N. Y. Tibbetts was transferred to Hawaii a second time in April, 1941. He was at Pearl Harbor on December 7, 1941. In February, 1942, he became Assistant Chief of Staff of the Seventh Air Force in the Pacific, in which capacity he served until December, 1944. He also commanded the VII Air Service Area Command during part of this period. In February 1945, General Tibbetts was assigned to Washington, D. C., in the Air Service Division, Directorate of Materiel. Prior to his assignment to WRAMA, he was at HQ SAC for five years.

In December, 1955, Gen Tibbetts was transferred to Wright-Patterson AFB as Comptroller of the Air Materiel Command. In July 1956, he became the first Director of Plans and Programming at Hq AMC, a position from which he retired in July 1958. General Tibbetts currently resides in Littleton, Colorado.

Major General Kingston E. Tibbetts
November 1952—December 1955

[3]Major General Thomas H. Chapman, whose biography is in the previous chapter, returned to serve as Commander from June 4, 1952—November 10, 1952.

Major General A. V. P. Anderson, Jr.
December 1955—October 1960

A. V. P. Anderson was born at Miles City, Montana, on November 8, 1905. He graduated from the United States Military Academy at West Point, New York in 1928.

During the early years of his service Anderson saw duty as a squadron adjutant, operations officer, squadron supply officer, flight commander, and squadron commander.

During World War II he flew B-24, B-25, and C-46 aircraft, and had 185 aerial combat missions to his credit. He was assigned as Commander of the Anti-Submarine Command, North Atlantic Division and as Chief of Staff of the 54th Troop Carrier Wing in the South Pacific.

In 1946 he became a member of the staff and faculty at the Air Command Staff School at Maxwell Field, Alabama. His next assignment was Chief of Air Section, U. S. Military Mission to Brazil.

General Anderson served as Deputy Commander of WRAMA from January 1, 1952 to March 1, 1954. He then went to Wright-Patterson AFB, Ohio as Director of Personnel and Support Operations at HQ AMC before his assignment as WRAMA Commander.

He is retired and lives in Shalimar, Florida.

Commanders, Fourteenth Air Force at Robins AFB

Maj. Gen. Charles E. Thomas, Jr.
July 1950—January 1955

Air National Guard and Air Force Reserve units mobilized in 1950 and 1951 for the Korean War. The command enlisted the first nonprior service reservists in March 1954 and later that year organized its first nonflying Air Force Reserve unit, an aerial port squadron at Greenville, South Carolina. General Thomas' biography is included in Chapter One.

Maj. Gen. George G. Finch
February 1955—July 1957

General Finch was born April 1, 1902 in Dade City, Florida. He is a graduate of the University of Georgia and a member of the bar in Georgia. He began his military career during World War I with the Air Corps. In 1926 he served with the Reserve Corps at Selfridge Field, Michigan and in 1940 he became a member of the Georgia Air National Guard. He served as Operations Officer of the Ninth Air Force in the Middle East, Executive Officer at an Army Air Base in Tullahoma, Tennessee, and commander of Army Air Bases in Bartow, Florida and Thomasville, Georgia. He became commander of the 54th Fighter Wing, Georgia ANG in 1946 and in 1948 was appointed chief, Air Force Division, National Guard Bureau in Washington, D. C.

In 1951 he went to Mitchel AFB, New York for CONAC and in 1952 was named Deputy Commander for National Guard Affairs. In 1953 General Finch went to Korea as the Senior Air Force Member of the United Nations Truce Negotiation Team. General Finch is retired and lives in Marietta, Georgia.

Maj. Gen. John W. Persons
August 1957—August 1959

General Persons was born September 19, 1899 in Montgomery, Alabama. He graduated from the Gulf Coast Military Academy in Gulfport, Mississippi, in 1915, and attended the Alabama Polytechnic Institute. He served with the Royal Flying Corp from 1917 to 1919. He was assigned to active duty in July 1928 and received his regular commission in the Air Corps on May 2, 1929. He served at Maxwell Field, Alabama, and Nichols Field and Clark Field, in the Philippines. In 1935 he was assigned to Bolling Field, D. C. and from there went to Air Corps Tactical School at Maxwell AFB, Alabama. In 1942 he was named commander of Tyndall Field, Florida. He also served at Kelly AFB, Texas, Elmendorf AFB, Alaska, and Randolph AFB, Texas.

Maj. Gen. Chester E. McCarty
October 1959—November 1960

General McCarty is a native of Pendleton, Oregon. He graduated from Northwestern College of Law in 1929. From 1930 to 1936 he served as assistant attorney general of Oregon, and in 1942, was elected as Oregon State Senator.

General McCarty was commissioned in the Air Corps in 1942. He was discharged after World War II, and was recalled to active duty in April 1951 as commander of the 403rd Troop Carrier Wing (Reserve) in Portland, Oregon. In 1954 he became commander of the Eighteenth Air Force and served as a commander in the Tactical Air Command for five years. General McCarty activated the Third Air Force Reserve Region in place of the discontinuing Fourteenth Air Force on July 1, 1960. General McCarty is retired and lives in Portland, Oregon.

Biographical Notes on the Fourteenth Air Force Commanders:

Maj. Gen. Ralph D. Steakley, October 1949—July 1950 (not pictured): General Steakley moved HQ Fourteenth Air Force to Robins from Orlando AFB, Florida, as that base was closing. General Steakley is retired and lives in Annapolis, Maryland.

Deputy Commanders, Warner Robins Air Materiel Area

Brig. General A. V. P. Anderson
January 1952—March 1954

Biographical Notes on the Deputy Commanders:

Brig. Gen. Stanley T. Wray, July 1954—December 1955 (not pictured): He is a 1932 graduate of the U. S. Military Academy and has held various positions including that of air inspector of the United States Air Forces in Europe (USAFE), Commander of the 7497th Airlift Wing, Germany, and chief of the Construction Division, Directorate of Installations, Washington, D.C. Gen. Wray left WRAMA to become Chief, Electronics Defense System Division of AMC's Directorate of Procurement and Production, New York City, in December 1955. He is retired and lives in Shalimar, FL.

Colonel Leroy Hudson, December 1955—June 1957 (not pictured): Colonel Hudson came to Robins in June, 1953. He later became Base Commander and Assistant for Programming. Col Hudson graduated from Texas A & M where he received a Bachelor of Science degree in Civil Engineering in 1927. On completion of his college training, Col Hudson entered military service in 1927 as a commissioned officer. Colonel Hudson's overseas assignments took him to Hawaii (1932-35) and Guam (Guam Air Depot from 1944-47). He received both the Air Medal and the Bronze Star. Col Hudson retired at Maywood Air Force Depot, CA after 31 years of active duty. His wife is the former Miss Helen Hutchins of Riverside, CA. They have three sons—Robert Gregory, Hugh Steven and Kent Leroy.

Col. Norman C. Spencer, June 1957—June 1960 (not pictured).

WRAMA Directors/Deputies

Col. Lloyd E. Johnson, Jr.
Director of Supply and Services
August 1951—July 1954

Col. Walter B. Packard
Director of Supply and Services
July 1954—March 1956

Col. B. J. Reilly
Director of Supply and Services
March 1956—December 1958

Col. Robert T. Peel
Director of Supply and Transportation
December 1958—March 1961

Mr. Paul K. Fain
Deputy Director of Supply and Transportation
November 1958—May 1971

Col. Arthur C. Perry
Director of Maintenance
July 1951—September 1952

Col. Henry O. Bordelon
Director of Maintenance Engineering
September 1952—January 1953

Col. Millard L. Haskin
Director of Maintenance Engineering
January 1953—April 1956

Col. J. V. O. Weaver
Director of Maintenance Engineering
April 1956—December 1957

Col. Robert J. Koster
Director of Maintenance Engineering
December 1957—December 1958

Col. Brendan J. Reilly
Director of Maintenance Engineering
December 1958—June 1959

Col. Walter F. Kutschera
Director of Maintenance Engineering
June 1959—June 1960

Mr. Elwood Maggard
Deputy Director of Maintenance
March 1953—September 1965

Col. Robert J. Koster
Director, Logistics System Management
November 1958—June 1961

Mr. Alonza Hassell Cotton
Deputy Director, Materiel Management
November 1958—January 1980

Col. Andrew Meulenberg
Director of Procurement and Production
April 1953—August 1954

Col. Paul I. Doty
Director of Procurement and Production
September 1954—August 1957

Col. Thomas O. Lawton, Jr.
Director of Procurement and Producton
August 1957—June 1961

Mr. W. W. Goetz
Deputy Director, Procurement and Production
April 1958—December 1972

Col. Walter R. Washburn, Jr.
Chief, Programming Office
December 1955—December 1958

Maj. James P. Smith
Chief, Plans and Management Office
January 1959—June 1961

Mr. Robert P. Schroeder
Deputy Chief, Plans and Programming Office
December 1955—December 1958

Mrs. Eleanor W. Richardson
Deputy Director, Plans and Programs
January 1959—December 1969

Col. Joseph S. Peddie
Base Commander
October 1957—September 1958

Lt. Col. Theodore H. Gorton
Comptroller
December 1951—August 1953

Col. Ernest B. Stevenson
Comptroller
September 1953—October 1954

Col. Laurence S. Fulwider
Comptroller
October 1954—April 1958

Col. Alfred G. Todd
Comptroller
July 1958—July 1962

Mr. Earl Barker
Deputy Comptroller
February 1952—November 1958

Mr. Carter Christie
Deputy Comptroller
November 1958—June 1961

Biographical Notes on WRAMA Directors and Deputies:

Col. Lloyd E. Johnson, Jr.: Col. Johnson was a leader in the complex job of decentralizing functions from Headquarters Air Materiel Command to the Air Materiel Areas. A Deputy for Requirements and Distribution was established within the Directorate at that time. Colonel Johnson is retired and lives in the Washington, D. C. area.

Col. Walter B. Packard: After his tour at Robins, Colonel Packard served as Director of Supply at AMFEA in Tachikawa, Japan, and as Director of Materiel Management at Rome AMA, NY and at Oklahoma City, OK. Following his retirement from the Air Force, he served as Program Manager for LTV. He has now retired and lives in Boerne, TX.

Col. B. J. Reilly: From his position as Director of Supply and Services, he transferred to become the first Director of Maintenance Engineering under the new reorganization. After his service at Robins, Colonel Reilly served as Director of Supply at Hq PACAF; graduated from the Industrial College of the Armed Forces; served as a civilian executive in the Defense Logistics Agency and in the Department of Transportation. Colonel Reilly is retired and lives in Locust Grove, VA.

Col. Robert T. Peel: Colonel Peel led in the WRAMA reorganization during the establishment of the Directorates of Supply and Transportation, which he headed, and of Logistics Support Management [later the Directorate of Materiel Management]. After his service at Robins, he served as Director of Supply at Mobile AMA, from which position he retired. He now lives in Spokane, WA.

Mr. Paul K. Fain: Mr. Fain was the first deputy of the new Directorate of Supply and Transportation, forming and leading in its successful establishment. He led in Quality Control, Production Control, and Grass Roots Improvement Program (GRIP) initiatives in the Directorate. He successfully enhanced the careers of the people who made up his work force through the "new man" concept, effectively using training programs to teach his employees a wider range of warehousing functions and creating a sense of identification with and worthiness of the distribution mission. Mr. Fain retired in May 1971; he resides in Warner Robins where he has been active in civic affairs.

Col. Arthur C. Perry: Under his leadership, a USAF-directed pilot automotive repair shop was established to serve as a center for testing improved industrial management methods in automotive repair. Colonel Perry is retired and lives in Ranch Sante Fe, CA.

Col. Henry O. Bordelon: The AMC decentralization program was implemented during his tour: the Command Workload Control Office was established with five branches (Aircraft, Equipment, Manufacturing, Vehicle and Powered Ground Equipment, and Command Contract Specifications) and the Technical Services Division set up to provide technical supervision of installation and maintenance. Colonel Bordelon is retired and lives in Boerne, TX.

Col. Millard L. Haskin: Under his leadership, the IRAN concept evolved, jet engines began to replace the conventional reciprocating engines, the TM-61 missile workload was assumed, and overhauling of "E" series fire control system began. Colonel Haskin served as head of the Pennsylvania Highway Department until recently. Colonel Haskin is retired and lives in California.

Col. J. V. O. Weaver: A major accomplishment during this period was the completion of Project High Flight in which some 2,800 new aircraft were inspected, winterized, safety checked, and flight tested. He is deceased.

Col. Robert J. Koster: Col. Koster became the first Director of what was later named the Directorate of Materiel Management, moving from his position as Director of Maintenance Engineering. Following his service at Robins, he taught at the college level in Texas. Colonel Koster is retired and lives in Stuart, FL.

Col. Walter F. Kutschera: During this period, Building 640 was modified (including clean rooms) for overhaul of MA-1 Air Weapon Control System. Electronics workload increased through support of B-52, B-58, and F-105 aircraft with AN/ASG-19/-21 Fire Control systems. Colonel Kutschera is retired and lives in Naples, FL.

Mr. Elwood Maggard: Under his leadership, the Directorate people worked on the F-86 Sabre, F-80 Shooting Star, F-84 Thunderstreak, F-100 Super Sabre, C-123 Provider, C-130 Hercules, C-47 Sky Train, B-57 Canberra, B-29 Super Fortress, and B-50 Super Bomber. Mr. Maggard died unexpectedly while serving as Deputy Director of Maintenance.

(*Mr. Cotton:* see Chapter One notes.)

Col. Andrew Meulenberg: Colonel Meulenberg retired from active service at Robins on August 21, 1954, accepting a position with Georgia-Lockheed Corporation.

Col. Paul I. Doty: Colonel Doty demonstrated outstanding managerial ability in placing large procurements for critical aircraft bombing and navigational systems, components, and supporting equipment. He retired from Robins.

Col. Thomas O. Lawton, Jr.: His concepts of management were responsible for WRAMA winning the Talbott Award two times in three years for excellence and efficiency. Colonel Lawton left WRAMA to go to the Dayton Air Force Depot and later became Commander of the Heath Operations Center at Newark, OH. He has now retired and lives in Florida.

Mr. Rush Gibbs, Deputy Director Procurement and Production, 1954—1958 (not pictured): Mr. Gibbs retired in early 1958.

Mr. W. W. Goetz: Mr. Goetz was instrumental in the directorate receiving the Brig. Gen. Nelson S. Talbott Award in 1960. Between July 1958 and December 1960 his efforts helped obtain competitive purchases of bearings for the Air Force. His "Vendor Program" resulted in a savings of $8 million in FY 1961 over the previous year. Mr. Goetz is retired and lives in Houston County near Centerville.

Col. Walter R. Washburn, Jr.: Colonel Washburn returned to Robins in 1968 as WRAMA Inspector General and retired here on January 31, 1970. He is currently living in Colorado Springs, CO.

Maj. James P. Smith: Major Smith, who was the first chief of the Plans and Mangement Office, moved to head the Plans and Programs Division. Major Smith retired in November 1966.

Mr. Robert P. Schroeder: Mr. Schroeder served as Mayor of the City of Warner Robins, where he is now retired and living.

Mrs. Eleanor W. Richardson: Mrs. Richardson was the highest rated woman employee in AFLC and the only female deputy on the WRAMA staff. Mrs. Richardson retired in 1969 to the Gulf Coast of Florida. She died in February 1982.

Col. Leroy Hudson, Base Commander, August 1953—June 1954 (not pictured): Colonel Hudson was the first Base Commander to be assigned after the function was separated from the Depot Commander in 1953.

Col. Wesley S. Calverley, Base Commander, June 1954—July 1955 (not pictured).

Col. Norman C. Spencer, Base Commander, September 1955—June 1957 (not pictured): Colonel Spencer later became the WRAMA Vice Commander. He retired as a Brigadier General and now lives in McLean, VA.

Col. Earl D. Duncan, Base Commander, September 1958—December 1961 (not pictured): The 4137th Strategic Wing (SAC) was established at Robins during Colonel Duncan's tenure.

Lt. Col. Earle C. Alley, Chief Personnel and Administration, January 1950—September 1950 (not pictured).

Col. Antonio N. Lagrippo, Chief, Personnel and Administration, September 1950—November 1951 and January 1952—August 1956 (not pictured): Colonel Lagrippo was reassigned to HQ SPAMA, Madrid, Spain.

Col. William T. Clayton, Chief, Personnel and Administration, August 1956—February 1957 (not pictured).

Col. James F. McClendon, Chief, Personnel and Administration, August 1957—July 1961 (not pictured): Colonel McClendon retired in 1961 and lives in Foley, AL.

Lt. Col. Lyman H. Brewster, Jr., Comptroller, March 1951—October 1951 (not pictured): Colonel Brewster was reassigned to the 75th Air Depot Wing at Kelly AFB, TX.

Lt. Col. Theodore H. Gorton: Colonel Gorton left Robins for an assignment at Ft. Mason, CA.

Col. Ernest B. Stevenson: Colonel Stevenson was reassigned to WRAMA Procurement and Production.

Col. Laurence S. Fulwider: Colonel Fulwider left Robins for an assignment in Japan.

Col. Alfred G. Todd: Colonel Todd left Robins for an assignment as Deputy Chief of Staff, Comptroller, HQ Fifth Air Force in Japan.

Mr. Earl Barker: During his term the Comptroller continued to grow and experienced many changes such as replacing manual posting with a mechanized process, separating budget and accounting functions, combining the Management Analysis office with the Plans and Programs office, and establishing a PR-MIPR control office. There was considerable increase in workload and manpower. The one area that expanded was the mechanized system for maintenance cost reports. The check writing process was mechanized and staggered pay days commenced. Mr. Barker is retired.

Mr. Carter Christie: During Mr. Christie's tour the field facilities expanded to include, first in 1959, a medium scale computer (Sperry-Rand UNIVAC File Computer) and then in 1960, a large scale computer. These expansions marked a major step forward in automated data processing. An innovative project in the Accounting and Finance area was to develop a direct deposit method to banks to eliminate the use of man-hours in payroll distribution. Mr. Christie contacted the local banks, obtained approval from the Treasury Department for a trial period and developed a procedure that was implemented and is now the direct deposit for all government agencies. Mr. Christie is retired.

A Portfolio of Aerial Photographs of Robins Air Force Base and the City of Warner Robins, 1937-1973

The Northern Half of the future Robins Air Force Base as it appeared on November 29, 1937.

The Northern Half of Robins Air Force Base as it appeared on March 4, 1949.

The Northern Half of Robins Air Force Base as it appeared on March 29, 1955.

The Northern Half of Robins Air Force Base as it appeared on February 6, 1966.

The Southern Half of the future Robins Air Force Base as it appeared on November 29, 1937.

The Southern Half of Robins Air Force Base as it appeared on March 4, 1949.

The Southern Half of Robins Air Force Base as it appeared on March 29, 1955.

The Southern Half of Robins Air Force Base as it appeared on February 6, 1966.

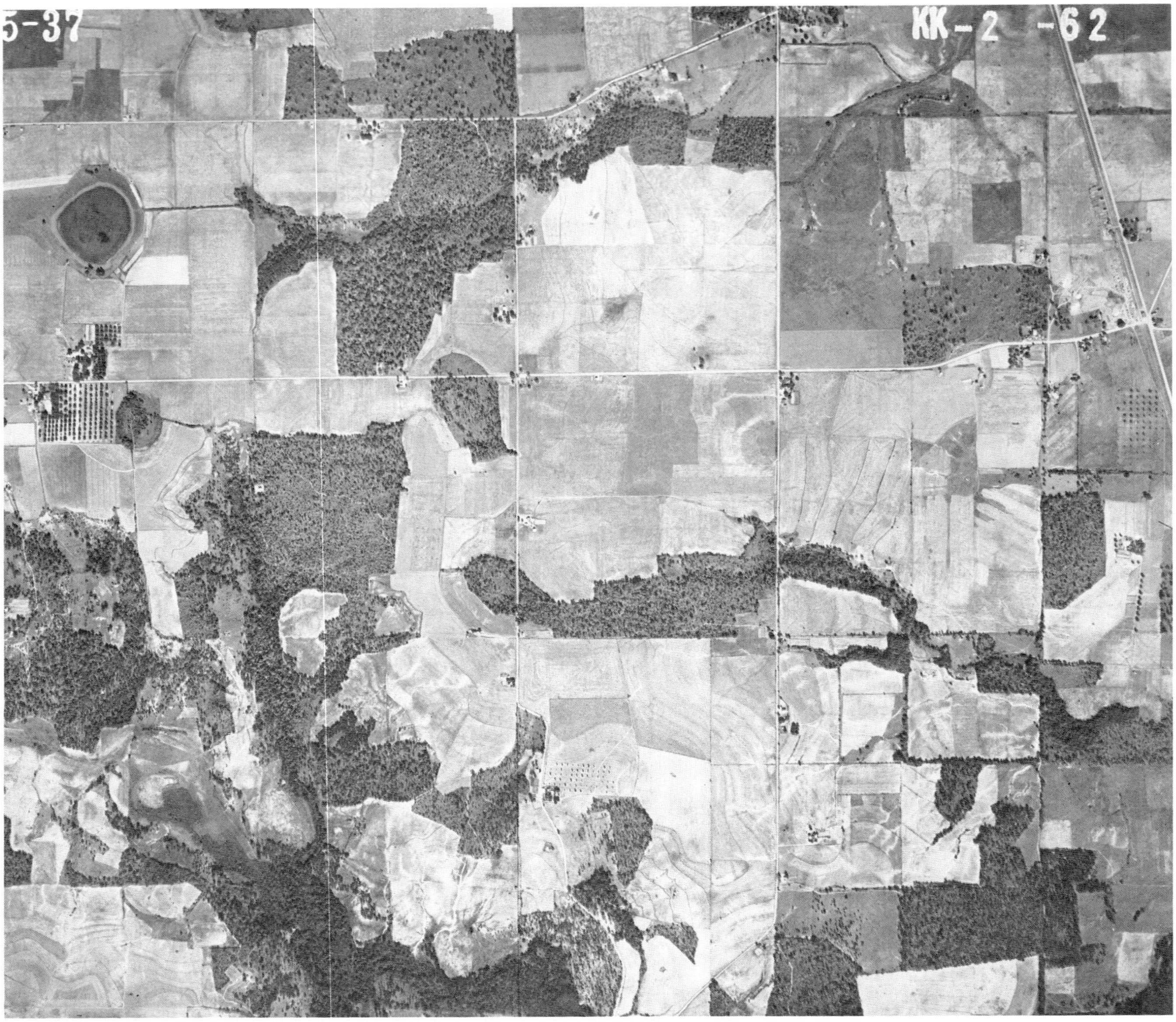

Part of the future City of Warner Robins as it appeared on October 25, 1937.

Part of the City of Warner Robins as it appeared on March 29, 1955.

Part of the City of Warner Robins as it appeared on January 7, 1960.

Part of the City of Warner Robins as it appeared on March 13, 1973.

Warner Robins Air Logistics Center and Robins Air Force Base . . .

The Largest Industrial Complex in Georgia

CHAPTER THREE: *The 1960s: Conflict Abroad and at Home*

1960s—Introduction

The decade of the 1960s was a watershed in American history; one of those periods when American foreign policy changed as presidents changed and as foreign problems emerged in new and difficult ways. Foreign policy in the 1950s was determined by John Foster Dulles. The conduct of foreign affairs focused on confrontation which critics said led to the brink of war, and defined the line beyond which the western alliance would not retreat. The communist advance was contained through area treaty organizations around the world; the nuclear response provided an umbrella under which American foreign policy could be actively advanced.

Beginning in the 1960s, a flexible response capability in addition to nuclear power characterized American diplomatic and military policy—meet crisis situations rapidly with conventional weaponry anywhere in the world. Such crises as Laos, Berlin, Cuba, the Dominican Republic and the naval ship "Pueblo" demonstrated American willingness to deal with potentially inflammatory situations at counterforce levels. Robins Air Force Base was part of that counterforce.

The decade was characterized by advances in technology. Transistors, printed circuit boards, and computers would change the nature of work being done at WRAMA. With these technological changes would come increased requirements for education and training; a highly skilled workforce would be necessary to maintain highly complex weapon systems.

Social issues motivated Americans to a more intense degree during this decade than at any previous time in the century. Robins people—Americans—were a part of or were affected by this movement.

"Let the word go forth from his time and place." President Kennedy and General Maxwell Taylor, Chairman of the Joint Chiefs of Staff, at Fort Stewart, Ga., in 1962.

It was a decade of hope and aspiration, a time of great national pride. Though hopes were unrealized and the crash from unfulfilled aspirations hit harder because of the height of the dream, there were positive endorsements, too; the Cuban Missile crisis and the space program, in particular, boosted American confidence.

Robins Air Force Base was also part of that new

generation of Americans of whom John Kennedy spoke in his Inaugural Address:

> Let the word go forth from this time and place, to friend and foe alike, that the torch has been passed to a new generation of Americans born in this century, tempered by war, disciplined by a hard and bitter peace, proud of our ancient heritage—and unwilling to witness or permit the slow undoing of those human rights to which this nation has always been committed, and to which we are committed today at home and around the world.
>
> Let every nation know, whether it wishes us well or ill, that we shall pay any price, bear any burden, meet any hardship, support any friend, oppose any foe to assure the survival and the success of liberty.

Troops departing from Robins AFB.

President Johnson arrives at Robins in 1964.

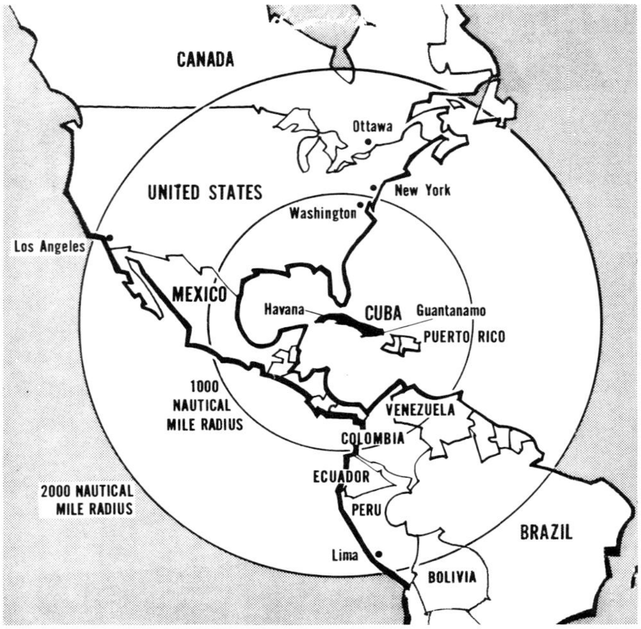

This map was used during the crisis to illustrate the range of the misiles based in Cuba.

Cuban Missile Crisis

During a press conference in August 1962, President John F. Kennedy announced that ". . . supplies and technicians of a rather intensive quantity are arriving in Cuba from the Communist Bloc . . ." He added that the Administration was "examining" the significance of this move. At the same news conference, a State Department official stated ". . . that since late July, at least twenty ships have brought three thousand to five thousand technicians to Cuba. Their cargoes were unloaded by their own crews at night and under guard." The Administration believed that Fidel Castro was receiving hardware to bolster Cuban air and coastal defenses.

During September and early October, the Soviets furnished repeated assurances that offensive weapons would not be placed in Cuba. On September 4, 1962, the President issued a statement that the United States would not tolerate the introduction of offensive missiles or nuclear weapons of any kind into Cuba. The United States continued to monitor the situation with U-2 flights over Cuba and maintained aerial reconnaissance over Soviet shipping in the Caribbean. The U-2 flight of October 14 obtained photographic evidence of the existence of the missile sites. On October 16, this information was presented to the President.

A week earlier, on October 9, WRAMA war planners began reviewing the applicable Operations Plan and making local preparations. On October 16, the same day President Kennedy was receiving photographic evidence, the WRAMA Command Post was conducting a briefing of WRAMA organizations outlining responsibilities in contingency plans for the Cuban/Floridian theater. A Comprehensive Plan was prepared and published on October 18 outlining specific actions including staffing, supplies necessary for operation, and transportation requirements for personnel moving to forward locations.

The prime function of WRAMA during the crisis, as always, was logistical support for the Air Force mission. Personnel and equipment were provided for SAC, MATS, TAC, ADC, and CONAC at the Florida bases of Homestead, MacDill, McCoy, Eglin, Tyndall, Patrick and Opalocka Air Station[1] and Sanford Naval Air Station. Personnel, here at Robins and at the various bases in Florida, were on twenty-four hour a day call and worked numerous hours of overtime in order to "keep .em flying."

Equipment provided to the various using agencies included such items as flood lights (to facilitate night work), tanks and pylons for F-100 and F-104 aircraft, LAU-3 missile launchers, tires for MATS aircraft, "C" band jamming capability oscillators for SAC, chaff,[2] GAR-8 Sidewinder air-to-air missiles, GAM-8A Bullpup air-to-surface missiles, napalm, rockets, 50 caliber ammunition, survival gear, and life preservers. President Kennedy spoke to the nation, informing them of the crisis, on October 22. AFLC had instructed WRAMA to have necessary materiel in place not later than October 20, two days earlier.

Work continued on aircraft modifications by Maintenance on C-54Ds for TAC and accelerated on SAC B-52s. WRAMA maintenance operations continued around the clock throughout the crisis.

The lack of technical orders in forward locations (to provide instructions on installation and maintenance procedures for the equipment arriving daily) was overcome by rushing these from WRAMA by the fastest means available.

On October 26, President Kennedy received a letter, evidently written by Premier Nikita S. Krushchev (without the assistance of the Politburo) which stated "If assurances were given that the President of the United States would not participate in an attack on Cuba and the blockade lifted, then the question of the removal or the destruction of the missile sites in Cuba then would be an entirely different question." The Politburo sent a message the following day offering to swap—we remove our missiles in Cuba; you remove yours in Turkey. The President elected to respond to the first letter; the Russians agreed and we stepped back from the brink of nuclear war.

On November 5, 1962, WRAMA's Command Post discontinued round-the-clock operations and returned to normal duties.

The success of WRAMA's efforts is best determined by those served. Brig. Gen. William G. Hipps, HQ TAC, made the following statements in his letter to Maj. Gen. William T. Hefley, WRAMA Commander:

> . . . essential shop equipment and weapons were arriving at the employment bases in support of the Tactical Air Command mission well in advance of the prescribed time limits. This effort on the part of your depot is one of which you and your personnel can be justly proud.

Technology

Advances in technology took a quantum leap in the 1960s. The move from tubes to transistors created drastic changes in the technology of the aerospace, communications and computer industries. "State-of-the-art" was a relative term as new innovation came about daily. Many outgrowths of the space program produced improvements in equipment managed or used by WRAMA. Lasers were virtually unknown at the beginning of the 1960s; however, by 1969 WRAMA was involved with various systems utilizing laser technology. Clean room facilities were constructed at WRAMA to provide organic repair capabilities for complex electronic systems.

In Maintenance work, the use of X-Rays, the electron beam welder and other new technological developments resulted in faster methods and better procedures. The use of non-destructive X-Rays permitted technicians to locate problems without disassembling equipment. The increased use of automatic test equipment allowed workers to perform numerous tasks in a minimum amount of time.

Another example of the advances, the secretary of 1969 worked under far better conditions than her counterpart in 1960. The secretary of 1960 typed on a manual typewriter (unless she was one of the lucky

[1]During the Cuban Missile Crisis, in order to establish additional facilities in Florida, Opalocka Air Station near Miami was activated. This experience of establishing an operational facility in a matter of days from a bare base became the cornerstone of the 4400 Mobility Support Flight (TAC), the first Gray Eagle activity, later renamed Harvest Eagle. Moved to WRAMA in the middle 1960s, this unit would play an important role in the war in Vietnam.

[2]Chaff is defined as electromagnetic radiation reflectors in the form of narrow metallic strips, used to create radar echoes for confusion purposes.

few on base having an electric), corrected her errors by hand, spent long periods of time trying to place long-distance calls, and typed numerous copies (perhaps occasionally utilizing the Thermofax machine). By 1969, the secretary could type an original and then run copies on reproduction machines, long distance calls could be placed on the AUTOVON lines faster, and correctable typewriters made typing much easier. All of these technological advances made her job simpler.

Computers went from first to third generation during the 1960s. The early large, bulky tube-type computers had been supplanted by transistorized, smaller models having a much greater capacity by the end of the decade. The third generation computers provided forecasting based on historical data, whereas the earlier models merely provided listings of those data. These computers were utilized for everything from processing paychecks to determining how many spare parts were needed in Southeast Asia.

Automatic Test Equipment allowed workers to test systems rapidly.

Working conditions for secretaries and support people improved with advances in office equipment.

Ronnie J. Fairfield calibrates a gyro. The gyro shop was built in the 1960s.

Improved computers made WRAMA's paperwork easier.

Autodin

AUTODIN provided improved communication by the use of computers. The system consisted of five switching centers, connecting 450 bases of the using commands, contractors, depots, and defense installations. This network transmitted and received computer tapes (punched or magnetic) and punched cards. By 1964 WRAMA's AUTODIN punch card terminal sent and received over four hundred thousand cards per month with an additional seventeen hundred teletypes received each month. The magnetic terminal received 22,800 messages per month in 1964 with one and one half million different items warehoused in thirty-nine different countries.

AUTODIN was an effective method for locating/relocating material rapidly. AUTODIN was extremely useful for multi-agency projects such as Red Horse and Bitterwine during the Vietnam War.

Shortage of Engineers

Rapid advances in technology during the 1960s resulted in more complicated weapon systems, advances in electronic warfare and increased reliance on computers. A shortage of engineering personnel became a recognized problem and received considerable attention. In November 1967, AFLC Commander Gen. Thomas P. Gerrity stated that "Failure to secure qualified engineering personnel in a timely manner in support of weapon systems is a critical limiting factor in an AMA's ability to take on additional workloads." The increased workloads during phase-up and throughout the war in Vietnam further pointed up the growing need for engineering personnel.

To meet those needs, WRAMA utilized two methods to attract new engineers, recruiting from colleges and hiring from private industry. College recruitment seldom succeeded as graduates accepted the higher starting salaries and promise of more rapid advancement which private industry offered. Recruiters from Robins visited colleges and universities all over the country but WRAMA received few acceptances. Private industry proved a better source for WRAMA. Engineers from the South or those who believed Civil Service had more to offer in the long run came to Robins. These engineers provided the majority of those hired in the 1960s.

Education

Improvement of the educational level and of educational opportunities became the focal issue of the 1960s. Courses established for employees to obtain high school equivalency certificates were well attended. Once an employee obtained a GED, the individual was encouraged to enroll in college classes or for technical training. Vocational courses upgraded the skills of employees and provided career enhancement opportunities. On-the-job training, special courses at Robins and at other bases and attendance at Dudley Hughes, the vocational technical school in Macon, provided specialized training.

To develop the technical skills needed at WRAMA, a three-year program to train workers for electronics, sheet-metal and aircraft repair—an apprentice program—was developed. The workstudy course provided WRAMA employees with theoretical class work and hands-on experience. The apprentices were selected by testing and received a starting salary of $1.98 per hour.[3] The graduates received excellent training both in mathematics and the area of specialization. The apprentice program also furnished an avenue for advancement of minorities into occupations where they previously had been underrepresented. This apprentice program graduated its first class in April 1969.

A continuing goal of WRAMA during the 1960s was to improve the educational level of its people.

Fred L. Shines makes a bolt on a machine lathe at Robins AFB. After more than forty-five years away from the classroom, Mr. Shines, a general machinist, returned to his textbooks at age sixty and passed the high school equivalency (GED) test.

Undaunted by the low level of success in recruiting college graduates in the engineering disciplines, officials pushed programs which upgraded existing employees. The major initiative was the establishment of an accredited four year degree granting institution for Robins Air Force Base, a task which occupied the efforts of successive commanders throughout the 1960s. Speeches by commanders to civic groups and the state legislature coupled with an intensive public relations campaign brought the issue to the attention of the public.

---

[3]1982 apprentices receive approximately $7 per hour.

Courses from the University of Georgia Center for Continuing Education, held at Warner Robins High School at night, allowed employees to complete some coursework on base. But the Board of Regents, the governing body of the state supported University System of Georgia, required a certain percentage of courses to be accomplished on campus and limited the number of courses which could be transferred. The worker wanting to earn a degree almost invariably had to take a leave of absence to reside on campus, a process few were able to follow. But the base tried to help.

Operation Campus, a project to encourage WRAMA employees to attend area colleges, (Middle Georgia, Mercer, etc.) provided transportation to the nearby campuses. Extensive publicity through the *Robins Rev-Up* included courses offered, registration instructions, and frequent articles about WRAMA employees working toward degrees. Encouragement of employees consisted of tuition assistance, special courses, tutoring and allowing employees to attend selected classes during working hours.

The announcement by the Board of Regents of a junior college for the Macon area further complicated Robins' situation. The junior college, located twenty-five miles from WRAMA, improved educational opportunities for Middle Georgians, but for the base was of little help in meeting WRAMA's needs for a school which could grant baccalaureate degrees.

Masters' Degrees are presented by Col. William Kelso to George Finison, left, and Wyman Priester Jr. on behalf of George Washington University.

The number of colleges within a hundred mile radius of WRAMA—Fort Valley State, Macon Junior, Wesleyan, Mercer, Middle Georgia, Georgia College, etc.—may have been a limiting factor in the eyes of the Board of Regents for the establishment of still another college, but the base remained convinced that the existing colleges could not meet the needs of WRAMA's employees. The students needed night classes which were affordable, located near Warner Robins, and which would allow them to complete their education without having to transfer courses or lose credits. The residency requirement remained the major problem for WRAMA employees.

Courses from the Air War College were offered to civilians as well as military, the University of Georgia increased the number and type of courses offered and the American University offered courses for a time at a Warner Robins Center of their District of Columbia institution. The fact remained that WRAMA needed its own degree granting institution.

Robins Resident Center

During the second week of June 1970, the Board of Regents of the University System of Georgia made the long awaited announcement—approval for Robins AFB to establish its own college degree-granting institution. The on-base facility, named the Robins Resident Center, could grant Bachelor degrees in Business Administration. The educational program was to be administered by Macon Junior College, Georgia College at Milledgeville, Fort Valley State College and Middle Georgia College. The center officially opened in Building 1676 in the "Dormitory Area" of Robins Air Force Base in September 1970, offering associate degrees in Business Administration from Macon Junior, in addition to the four year degree from Georgia College.

After years of hard work, this sign could be erected—Robins had its own four year college.

The Resident Center was established solely for the use of Robins civilian employees, assigned military and their dependents. WRAMA provided the facilities, maintenance, and in many cases, tuition assistance with the colleges furnishing the professors' salaries.

This event finalized years of work by four successive AMA Commanders, General Hefley, Maj. Gen. Earl C. Hedlund, Maj. Gen. Francis C. Gideon, and Maj. Gen. A. J. Beck. They had focused the attention of politicians and educators on the educational needs of Robins Air Force Base through speeches to the state legislature, tours of WRAMA's facilities, and by getting local civic groups involved in the push for a four-year degree granting institution.

WRAMA's employees had demonstrated their desire for self-improvement by the number of participants in adult high school courses and in the college programs previously available. When the Resident Center opened in September of 1970, seven hundred people registered for the first quarter. Since that time, the Resident Center has maintained an average attendence of over 585 undergraduate students. From September 1970 to the present, 813 undergraduate (associate and bachelor) degrees have been awarded.

The educational level of the workforce by 1981 showed the results of efforts begun in the 1960s: 7.5 percent had less than a high school education (as compared to 43.3 percent twenty years earlier), 39.5 percent with high school diplomas (versus 45.2 percent), 41.4 percent had some college (as compared to 7.8 percent), and 11.6 with college degrees (as compared to 3.7 percent).

Efforts by WRAMA officials and community leaders had focused on securing a four-year college (with emphasis on hard mathematics and sciences) for Robins Air Force Base. The establishment of the Robins Resident Center was a beginning. The push continues today.

Lab equipment used in an electronics class at Robins Resident Center.

Robins Graduate Center

The graduate program (begun in 1967) provided a means for college graduates, engineers, and the bright young Air Force lieutenants to continue their education by obtaining an advanced degree. WRAMA could attract more college graduates and engineers if a means were available for them to further their education. A consortium of the University of Georgia (to provide Master's in Business Administration) and Georgia Tech (for advanced engineering degrees) taught these classes on base. In 1977, Georgia College replaced the University of Georgia and a course in public administration was added to the curriculum. The Graduate Center, limited to employees, military and dependents, has granted 571 master's degrees since opening.

Cost Reduction Program

In 1963, the Cost Reduction Program was implemented. This program was designed to improve procurement and logistics management and to reduce operating costs without impairing mission performance. The Air Force Logistics Command established a six-point objective: (1) buy only what was needed, (2) buy at the lowest price, (3) reduce operating costs, (4) promote cost awareness at every level, (5) improve AFLC's public image, and (6) use the results as a tool for measuring management effectiveness. Air Force officials said that this program was to reach the worker on the line and no task was to be considered outside the program's reach.

WRAMA finished first among the AMAs several times in the 1960s. General Beck and Colonel Gafford show the Cost Reduction Flag.

WRAMA Employees participation in Project Gold Rush was encouraged by signs such as these above and below.

In "Project Gold Rush," WRAMA employees were encouraged to "dig for gold" (develop cost reduction ideas) and to "stake their claims" (file the proper paperwork). WRAMA carried the symbol a step further with the monthly presentation of the "Gold Nugget Award" (actually a piece of Georgia marble painted gold) to the directorate showing the greatest cost reductions. Project Gold Rush became a frequent topic for discussion at the Commander's staff meetings.

AFLC assigned goals for each AMA; WRAMA achieved its goals in each of the years 1965-1969. At the end of the first year of the program, the Air Force cited AFLC for achieving eighty percent of the entire Air Force cost reduction savings. AFLC and WRAMA received letters of commendation from Secretary of Defense Robert McNamara,[4] Air Force Chief of Staff General Curtis LeMay and Secretary of the Air Force Eugene Zuckert. In his State of the Command Address, January 8, 1964, General Hedlund summed up the goals of the program: ". . . each government employee is responsible, not only for doing his assigned job to the best of his ability, but also for devising and proposing ways to improve his performance . . . We must put our imaginations to work to find a better, quicker, less expensive way to do our job."

[4]Robert McNamara served as Secretary of Defense under Presidents Kennedy and Johnson.

Project Count

In 1960, in order to obtain more accurate information for budget and buying programs, the Air Force conducted a worldwide inventory. This inventory, known as Project Count, was a fence-to-fence project covering one and a half million different items in approximately one hundred million square feet of storage space in bases all over the world. Previous inventories were conducted by stock class; however, this effort concentrated on a total computation of all assets.

During the inventory, all WRAMA prime stock classes were closed (routine requests were held for a few days) except for high priority requisitions. A number of personnel were lent to the Directorate of Supply and Transportation to participate in the count. The use of automatic data processing equipment allowed faster tallying of results, better reporting methods, and aided in verifying the accuracy of the inventory. The project officer was the Director of Supply and Transportation and the Quality Control Office was responsible for maintaining the accuracy of the project.

Project Count results revealed that WRAMA had six hundred thousand line items stored in 1.2 million locations. WRAMA also stored items for other AMA's and had items at 12 off-base areas. As a result of Project Count, new annual inventory procedures were instituted to make stock class inventories more accurate.

Zero Defects Program

The Zero Defects Program, also part of the overall Cost Reduction Program, was developed at the Martin Plant in Orlando, Florida, in the early 1960s and adopted by several other corporations. The Defense Department was impressed with the program and adopted it in 1964. The SAC wing at Robins was one of the many sites testing the program in the early 1960s.

For WRAMA, implementation came with the arrival of Captain Zero, a "Superman" type symbol in appropriate attire who appeared at large Zero Defects rallies held in work areas on base. The program encouraged workers to strive for error free performance in every task—get the job done right the first time. Several directorates reported that defective workmanship had been reduced from thirty to fifty percent during 1965. The program sought to make each employee aware of the importance of his or her own role in reducing waste and increasing productivity. Employees were awarded with buttons, plaques, and certificates for efficiency.

Captain Zero obviously exhorted WRAMA's workers to do the job right the first time.

But it didn't always work—"perfection" remained elusive.

Value Engineering

Another facet of the Cost Reduction Program acquired from private industry was Value Engineering. The program was introduced to Robins in late 1964 by a series of training seminars conducted with personnel from Materiel Management, Maintenance, Procurement, Supply and Transportation, and the Base Commander's staff. Value Engineering encouraged the design of equipment to strive for mission accomplishment without frills, using the least expensive materials and processes to get the job done. AFLC was primarily interested in Value Engineering as a means of cost-conscious redesign and reengineering of existing equipment. At Robins, formation of study groups composed of representatives of the participating directorates reviewed items with Value Engineering potential, reviewed planned actions and checked progress periodically, and followed through until savings were validated and reported.

Value Engineering consisted of three phases: organic (work performed at Robins resulted in greatest savings), contractual (the contractor provided Value Engineering Change Proposals with savings shared with the contractor), and funded (the contractor was paid for Value Engineering studies).

Employees developing Value Engineering ideas which were implemented received cash awards as in the Suggestion Program. (The Suggestion Program was and is for ideas not job related whereas Value Engineering can be for job related ideas.)

HQ AFLC set WRAMA's goals at between $8.5 million in 1967 (1967 was the first year of validated reporting to AFLC) to $33.25 million in 1981 and with the exception of 1968 and 1969, WRAMA exceeded these goals.

Procurement Methods

The cost-plus-fixed-fee contracts, prevalent in the 1950s, came under scrutiny by Secretary McNamara. He felt these contracts were inefficient as they guaranteed contractors a profit regardless of their performance. The firm-fixed-price contract, preferred by McNamara, encouraged the government to prepare more precise Statements of Work and encouraged contractors to meet contract dates within their budgets. Incentive contracting, a compromise procedural method suitable for development of new systems, had built-in bonus or penalty clauses which provided incentives for contractors to bring the contract in under cost and ahead of time.

By 1968, when McNamara resigned, fixed price contracts represented seventy-eight percent of defense procurement (as compared to forty-three percent in 1960).

These procurement methods proved a challenge to corporations. McNamara's concern was to obtain a given level of military effectiveness at the least possible cost. Quality was not to be sacrificed; the system procured must meet the mission requirements. Many of WRAMA's systems, modifications, and spares were purchased by these new methods—including the F-15.

Phase-Up

Secretary McNamara announced on November 19, 1964 that eighty bases were to be consolidated, reduced or discontinued including three Air Materiel Areas: Mobile, Alabama; San Bernardino, California; and Middletown, Pennsylvannia.[5] An AFLC 1964 study showed the eight AMA's maintenance facilities usage was sixty-five percent based on a forty-hour, five-day week and the study determined that by using only five AMA's a rate of eighty-five percent could be achieved with an annual savings of $86,000,000. Selection of the bases to be closed was based on type of mission, the ease of moving the mission, and the condition of facilities. WRAMA's clean rooms and other repair facilities for "black box" maintenance were major factors in Robins not being selected for phase out.

WRAMA representatives Chet Herlin and Dave Davis went to Middletown to recruit transferring workers and to learn about the functions being sent to WRAMA.

[5]Rome Air Materiel Area, New York had already been phased out early in 1964.

General Hedlund established an orientation program to urge the Mobile and Middletown[6] personnel to relocate at Robins. Teams composed of base personnel and local community leaders visited both bases. Community representatives discussed housing, schools, churches and recreation—emphasizing the fact that people from all over the United States lived in Middle Georgia. The base employees stressed the need for these people to follow their functions. There was an effort to put the best foot forward and lead the command in relocations in marked contrast to the poor relocation rates of the 1950s. Teams from Middletown and Mobile visited WRAMA and reported back to their co-workers.

WRAMA's efforts paid off in a higher acceptance rate from functional transfers than any other AMA. Many employees elected to remain in their respective areas, however, and accepted positions with the Navy, Army, and private industry. Even though thirty-five percent of the workers transferred, WRAMA was left with a shortage of trained personnel for the increased workload.

The items transferred to WRAMA included missiles, gyros, bomb-navigation equipment, parachute and flight clothing maintenance, helicopters, O-1 and P2-E observation aircraft, supply distribution for vehicle spares, cargo aircraft (C-46, C-119, C-123), among other items—most of which played an important role in the Vietnam war.

The relocation of equipment and records was a vast undertaking. In one week, WRAMA moved 154,147 pounds of records and equipment from Middletown and Mobile without losing a single carton. Instructions were given to Mobile and Middletown employees on packing everything from office supplies (paper clips, pens, etc.) to missiles. In most cases it was necessary to maintain the functions throughout the transfer. The parachute shop was dismantled, shipped to WRAMA, reassembled and produced the first parachute on schedule.

During the period of the initial planning for phase-up (late 1964-early 1965), the international situation once again would affect work at WRAMA. Ten thousand miles away, the United States would once again become a part of a battle in Asia: The war in Vietnam became front page news.

Office of the Placement Coordinator, Zone 2 and Atlantic Theatre

The Department of Defense has earned a reputation as a considerate employer with concern for each individual employee. During the 1960s when many military installations were declared surplus and deactivated and others had significant mission reduction, large numbers of civilian employees were threatened with unemployment. In order to retain the valuable skills of these persons and to enable them to continue their careers, the Department of Defense Stability of Civilian Employment was established in 1964. This program refers adversely affected employees to Defense activities having appropriate vacancies and mandates selection. Since 1964 more than seventy-seven thousand persons have been placed through this system.

Administration of this and other civilian personnel programs that are Department of Defense-wide in scope is through four zone and three overseas theater coordinator offices. Robins Air Force Base was selected as the site for a combination Zone 2/Atlantic Theater Placement Coordinator's Office and Mr. Lee C. Lingelbach was named to the top position, a job that he has held continuously since 1964. Mr. Lingelbach and his deputy, Mr. Gene Pollard, have responsibility for approximately 252,000 jobs at 186 locations in the 13 southeastern and southwestern states and 9 activities in the Atlantic Theater. The Zone 2/Atlantic Theater Placement Coordinator represents the Assistant Secretary of Defense (Manpower, Reserve Affairs, and Logistics) and works with local government bodies, Chambers of Commerce, and community action organizations in areas affected by base closures or large scale reduction in force.

The parachute shop was disassembled and moved from Middletown to Robins. It maintained its schedule during the transition.

[6]San Bernardino phase-out did not greatly affect the WRAMA mission as Robins received none of their functions and only a few of their people.

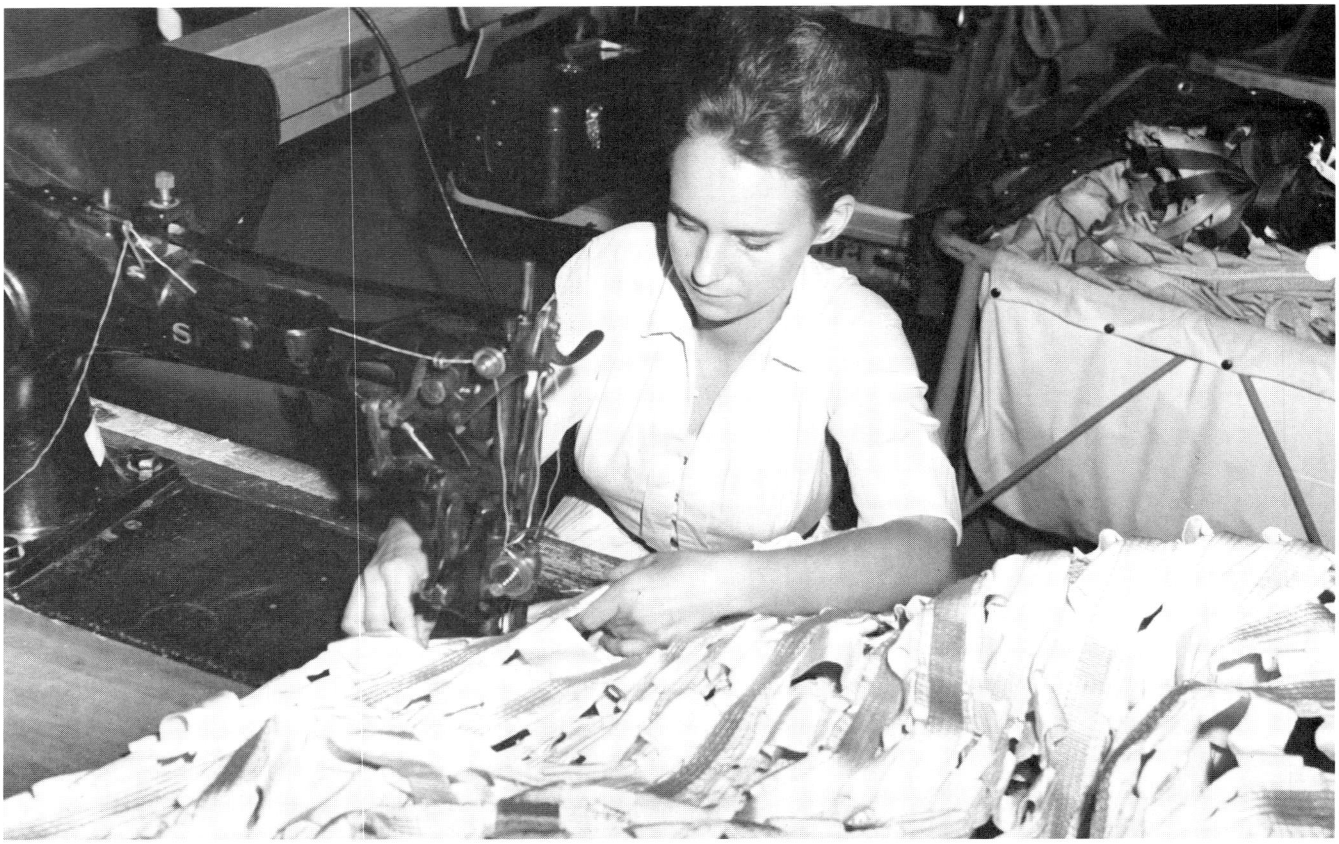

New jobs from the phase-up provided new career opportunities at Robins. Burma McGlamry is in training in the parachute shop in August 1966.

AFLC Management Engineering Team (MET), Detachment 6, 3025 Management Engineering Squadron

The AFLC MET began operation at Robins Air Force Base on January 12, 1964, as the AFLC Manpower Program Validation Branch, Detachment 23, 2750th Air Base Wing, with Mr. William H. Carter as chief.

The objective of the Manpower Validation Program (MVP) team was to systematically increase work center effectiveness, develop and maintain accurate work center manpower standards and recommend solutions to management problems.

The original emphasis of the MVP was on work center manpower standards; however, as the program progressed, there was additional emphasis on detailed and comprehensive management improvement. Consistent with this change, the title of the program was changed, in August 1964, to Management Engineering Program (MEP).

In December 1966, the AFLC Management Engineering Team (MET) was formed by bringing together the Management Engineering Program Team with the former WRAMA Manpower and Organization Divison. The organization consisted of the Management Engineering Office and the Manpower Office. MET responsibilities included the AFLC MEP, review and audit of labor standards, workload data and other factors developed by local industrial engineering activities; manpower program development and allocation of manpower authorizations, supervisor and high grade position control; management of overhire and overtime usage; and accomplishment of engineered organization studies.

The MET influence extended to the Allied Nations in the late 1960s and early 1970s. Foreign military students were trained by the local MET to apply MEP techniques in their native countries.

The Office of Management and Budget (OMB) placed a major emphasis on cost-effective operations as the MET entered the decade of the 1980s. The Government's general policy of reliance on the private sector for goods and services recognized that certain functions were inherently governmental in nature and must be performed by Government personnel, and that relative cost must be given appropriate consideration in decisions between in-house performance and reliance on private commercial sources. The MET is responsible for the government bid on a cost comparison study of a function identified as a candidate for contracting out.

Vietnam

Vietnam's history was one of frequent foreign intervention—the annexation by China in 111 B.C. is an early example. In the early Nineteenth Century western contact began, as it did over the world, with trade and missionaries. French domination was firmly established by the end of the century. The Japanese invaded Vietnam in 1940, but the country reverted to French control after the Japanese surrender. Led by Ho Chi Minh, the Vietnamese declared independence; the French-Vietminh war of 1946-1954 then ensued.

During this conflict, on July 1, 1949, the French established a government for Vietnam with the capitol at Saigon; the United States recognized that state on February 7, 1950. To assist the French, President Truman sent a 35-man Military Assistance Advisory Group to Vietnam in 1950. The French involvement ended when the Geneva Conference on Indo-China in 1954 divided the country at the 17th parallel. North Vietnam, for the most part, had been won from the French by Ho Chi Minh and his forces. South Vietnam was the surviving remnant of the Vietnamese government which the French had established. When the two Vietnams could not agree on an election to reunite the country, (as provided for at the Geneva Conference) North Vietnam sought to unite it by force of arms.

During the Eisenhower years, guerrilla activity by the Vietcong (Vietnamese communists) led to the increase of American advisors from the original number of 35 to 685 by 1960. With the change of administrations, President Kennedy reaffirmed American support in his statement of December 14, 1961, "We are prepared to help the Republic of Vietnam to protect its independence."

General Hedlund reviewing map of Vietnam with Civic leaders: Left to right: Macon Mayor B. F. Merritt; General Hedlund; Dr. Dan Callahan, local physician; and Mr. Peyton Anderson, Publisher of The Macon Telegraph.

American involvement continued to grow. "Assistance" became "escalation" and "advisors" became "soldiers" following the reported attack on U. S. destroyers in the Gulf of Tonkin. The Gulf of Tonkin Resolution passed by Congress on August 7, 1964 gave President Johnson the power ". . . . to take all necessary measures to repel any armed attack against the forces of the United States. . . ." The necessary measures continued to grow.

General Hedlund with Vietnamese Air Force Officers at Robins.

Until 1967 the vast majority of the American people supported the war. Belief that the end was in sight (McNamara stated he could see the light at the end of the tunnel) and that the war would end rapidly became disillusionment—with body counts, the corrupt Saigon government, and the nightly scenes of destruction (Vietnam was the first "televised" war). The McCarthy "peace" campaign, President Johnson's decision not to seek reelection, Kent State, and draft evasion all were evidence of developing national unrest. Ending a losing situation was becoming a viable alternative.

The first Nixon administration was ending before the Paris Peace Treaty was signed by Henry Kissinger and Le Duc Tho; the last American troops withdrew from Vietnam and in January 1973 the prisoners of war came home. In 1975, the light at the end of the tunnel went out and the tunnel itself collapsed—South Vietnam fell.

Robins Air Force Base had been a major contributor to the Vietnamese experience. It was a time during which the population of the base reached its highest levels since the end of World War II and its impact on logistical support of the fighting commands tied the home front to the front lines as never before.

Vehicles/Spare Parts Support

In an early effort to gain efficiencies through single manager support, the General Services Administration had been designated as the logistical source for government vehicles. The GSA required offices on the west coast at Seattle and San Francisco to provide vehicle support through local purchase of spares for government vehicles in the Pacific.

With the quick increase in American participation in Vietnam, vehicle parts requests suddenly inundated GSA to such a point that vehicle support almost came to a halt. Faced with an intolerable situation, the Secretary of the Air Force decided in 1966 to withdraw from the GSA system and establish an Air Force resupply system with WRAMA in charge.

Things were a mess. Spares were being ordered by manufacturer's stock number and there were innumerable common parts being purchased on an individual basis.

WRAMA rapidly developed SEAVICS (Southeast Asia Vehicle Inventory Control System) during the summer of 1966 to cross reference parts and to facilitate ordering, storage, and shipment of spares. Procurement and Materiel Management established rapid ordering procedures to reduce the backlog of materiel requested.

Once WRAMA had SEAVICS functioning properly with requisitions being processed in a timely manner, and "the bugs worked out of the system," the Air Force decided in 1972 to transfer vehicle spares support to an element of the Defense Logistics Agency in Ohio. During the latter years of the war, the Defense Logistics Agency supported vehicle spares using the methods developed by WRAMA.

Vehicles such as this were supported by WRAMA during the war in Vietnam.

Build-Up

Air Force personnel arriving in South Vietnam in the early 1960s found maintenance facilities inadequate—bases which had been used by the French or built under the Military Assistance Program. The buildup of 1964-1965 led to greater requirements for construction in areas considered inaccessible to contractors. The number of Air Force personnel also increased substantially during 1965, resulting in requirements for additonal housing, wells, roads, office buildings, maintenance facilities, and runways. To meet these needs, the Air Force established two organizations—Prime BEEF (temporary) and Red Horse (permanent).

Prime BEEF team preparing a pallet in exercises at Robins.

Prime BEEF

Prime BEEF (Base Engineering Emergency Force) teams consisted of approximately twenty-five man teams composed of various specialists (electricians, welders, carpenters, etc.) deployed on TDY to repair runways, build revetments to protect aircraft, and perform other construction duties. These units were generally deployed for 120 days, carrying enough equipment and supplies to perform their missions, with resupply from Clark Air Base, Philippines, if needed. The Prime BEEF teams still function as part of the Air Force today.

Loaded Pallets such as those shown here were shipped to Vietnam as part of WRAMA's efforts to support the war. This picture of 2853CES Prime BEEF Team taken in July 1982 shows team members preparing straps for a pallet.

Red Horse

In 1965, the Air Force determined that a more permanent organization was required to provide an effective rapid response for minor construction and to upgrade existing airfields. The first two Red Horse squadrons (Rapid Engineer Deployable Heavy Operations Repair Squadrons, Engineer) received orders to deploy in January and February 1966. AFLC designated WRAMA as control point for Red Horse based on WRAMA's vehicle management. A series of meetings in October 1965 established the equipment requirements and coordinated the activities of the many agencies involved.

For Red Horse, the Air Force designated Gulfport, Mississippi, as the Port of Embarkation. WRAMA furnished a liaison representative to Gulfport to facilitate this operation. WRAMA obtained an agreement with the Navy to paint the vehicles, establish an assembly point, provide vehicle repairs, apply corrosion treatment, and load vehicles on receipt of movement orders.

Since the two squadrons were scheduled to leave in January and February 1966, a great deal of work was accomplished in a short period of time. The Data Products Division (Materiel Management) kept flowtime control of each item in the shipments and maintained progress records for each agency. Item managers located equipment at other AMAs and arranged for transportation. Procurement expedited ordering (some items normally requiring a year for delivery were arriving at Gulfport in less than 90 days).

Vehicles from WRAMA being loaded for Vietnam at Gulfport, Mississippi during Project Red Horse.

Under WRAMA's direction the Army modified vans to serve as mobile shops. Supply provided special packaging and palletizing which would protect the supplies from corrosion and allow easier unloading at the final destination. WRAMA provided the same services as in normal conditions but because of the time factor, the number of agencies, and variety of contractors involved in the project, coordination of effort was a key factor.

The first two squadrons departed on January 9, and February 7, 1966—one for Phan Rang and the other for Cam Ranh Bay. The Air Force determined that additional squadrons would be required before the equipment was assembled for the first two squadrons. For several years, Red Horse was a continuing project at Robins.

Bitterwine

Air operations against North Vietnam began in 1965, resulting in the deployment of additional men and planes to Southeast Asia.[7] The Air Force had to establish new bases and expand the existing ones. A new project, "Bitterwine," under the direction of AFLC supplied the materiel. HQ AFLC designated Sacramento Air Materiel Area as the control point for Project Bitterwine and instructed the AMA's to provide their full support for this project. Before Bitterwine got off the ground, the tempo of United States involvement stepped up and additional bases were required—an increase from four to twenty-two.

Prior to 1965, Air Force personnel served on thirty-day TDY rotation tours to Vietnam; however, the Air Force decided to transfer certain squadrons permanently to Southeast Asia. Rapid changes in methods to meet rapidly changing needs in the Southeast Asia conflict made planning difficult.

AFLC's responsibilities consisted of determining requirements, procuring, packaging, and shipping the necessary items. Bitterwine accomplished the processing of "packages" rather than an individual piece of equipment. Each package consisted of specialized materiel to support a specific function (food service, weapon system, etc.) necessary to set up the new bases.

Building the bases consisted of three phases: "Gray Eagle" Kits (tent facilities); prefabricated buildings;

---

[7]Geographically, Southeast Asia included Laos, Thailand, Vietnam, Cambodia, and Indonesia, but the Air Force used this term to include all locations in the Pacific area assigned to support the war in Vietnam. Such bases as Misawa in Northern Japan, Kadena in Okinawa, Clark in the Philippines, and several bases in Thailand all served as part of the ten thousand mile pipeline which linked South Vietnam with the Air Materiel Areas.

and final construction (housing, runways, support facilities, etc.) The lessons learned from the Cuban Missile Crisis were carried on by Gray Eagle (TAC) support throughout Vietnam.

WRAMA established a control point for Bitterwine and each affected directorate had control officers and alternates appointed. All Bitterwine requests received priority handling and to ensure serviceability on arrival, WRAMA implemented one hundred percent inspection procedures. Requests went through Sacramento and were then forwarded by them to the appropriate AMA.

The deployment of entire MAC, SAC, and TAC wings created requirements for spare parts and special equipment. WRAMA also supported commodities for the using commands such as bomb/navigation systems, fire control systems, missile launchers, tires and communication/navigation equipment—all of which required spare parts, modification, and maintenance. WRAMA's workload included development, assembly, and shipping of "packages" in pre-warehoused bins for two C-130 wings, a squadron composed of C-123s (a small cargo plane), U-6s and UH-1Fs (helicopters) and AC-47s (gunships), twelve missile packages, and a CH-3C (helicopter) squadron.

WRAMA obtained the materiel from on-hand assets, procurement, redistribution from bases being closed, and from operational bases (with the exception of the National Guard, Reserves, and NATO bases). WRAMA developed the special packages, performed funding action, and established warehousing, packaging, packing, and transport facilities. The vast amount of materiel gathered (each item with a different stock number), the number of agencies involved, and the critical time factor led to a computerized reporting system. This system reported which requisitions were outstanding, those having incorrect stock numbers, and those which showed increases or decreases in requirements.

Each of the Air Materiel Areas used various forms of transportation—air, rail, truck, or parcel post—to get the items to Sacramento. From California to Southeast Asia, most packages were transported by ship as an economy measure. Changes in requirements or squadron activation dates created the need for airlift; however, in 1965, 95.5 percent of shipments used water transport. Frequently it was possible to load all equipment needed or required to establish a base on a single ship.

By the end of 1966, over one hundred thousand pounds of equipment (twenty-two percent from WRAMA) had been sent to Southeast Asia—making Project Bitterwine one of the largest logistics efforts since the Korean War.

Single wall inflatable shelter erected at Cam Ranh Bay.

A part of the Harvest Eagle (formerly Gray Eagle) Air Transportable Housekeeping Package which can support 1,100 people under bare base conditions anywhere in the world.

RAMS/RASS/RATS

Not only military special purpose teams, but also civilian volunteer maintenance, transportation, and supply teams were furnished by WRAMA to the Southeast Asia theater.

"Rapid Area Maintenance" (RAM) teams repaired crash or battle damaged aircraft requiring work beyond the capabilities of field maintenance personnel. The aircraft (repaired on site, prepared for one-time flight to Air Force or contractor facilities, or

RAM Team members in Vietnam fixed seventy-two holes in pressure compartments, 158 fuselage holes, and 271 nicks and dents on this WRAMA managed B-57.

This group of civilian employees assigned to WRAMA were recognized for their temporary duty efforts in Southeast Asia. They served on RAM, RATS, and RASS teams or were WSLOs. The "Civilian Service in Combat Areas" pin was presented each individual by General Gideon, WRAMA Commander. Pictured above are David F. Addison, Willard Adkins, Mack E. Bowman, Willis Bradley, Leo C. Casey, Franklin D. Pritchett, Hubert P. Byrd, James R. Hester, Jr., Harry J. Mason, Arthur P. Melton III, Leo E. Reese, Walker D. Taylor, E. Hollis Whitley, Algie Collins, Jr., Robert L. Haddock, James W. Hopkins, Bobby E. Jones, Claude E. Renfroe, Jr., Milton T. Smith, Aaris R. Ussery, Oscar W. Woodward, Jr., and Fred Malone. Also Glen H. Arnold, Theron P. Bramblett, Jack R. Young, Frank Brantley, Jaspar L. Browning, Kenneth L. Donelson, Ernest Q. Green, Eugene Morris, Jaudon Thompson, Ralph L. Wells, William L. Carswell, Burke D. Giles, Jr., Alton E. Glenn, Thomas B. Stinson, and Preston E. Teal. Also, Ronald Bush, Robert J. Cranford, John Welton Barrs, Jr., Oscar D. Lee, Jr., Ray Fendley, Jack L. Sutton, Charles A. Rodgers, Jr., Charles H. Kezar, John L. Mays, Rembert D. Owens, Henry J. Saturday, Cecil H. McDaniel, Leroy Morris, Frederick C. Poole, and Gerald B. Reeves.

disassembled for spares) included F-4 and century fighters, B-52 bombers, C-130, C-124, C-141 cargo aircraft, and helicopters.

RAM team members frequently worked and lived under combat conditions. In one instance, a WRAMA RAM team at An Hoa replaced a wing on a C-130 only to have a Viet Cong mortar attack destroy the opposite wing. Ten days later, after the second wing was replaced, the C-130 flew out during yet another mortar attack. RAM teams worked at numerous sites in Vietnam.

To improve cargo handling, "Rapid Area Supply Support" (RASS) teams from Robins and the other AMAs moved to remote areas to establish inventory, accounting, storage and issue procedures. RASS teams were composed of supply logistics specialists, warehousemen, packaging specialists, traffic supervisors, and packers. Each AMA provided volunteers and teams consisted of personnel from each AMA.

Rapid unloading of equipment at docks in Vietnam remained a problem throughout the war; difficulties existed in getting the Vietnamese to do anything rapidly. Rapid Transportation Support teams (RATS) assisted Pacific Air Force personnel at both aerial and water ports.

WRAMA Volunteers

In addition to aircraft, weapon systems and spare parts, WRAMA personnel volunteered to serve in Southeast Asia. Civilian employees served overseas as weapon system logistics officers, supply personnel, maintenance workers, etc., on either a temporary duty or one-to-three year permanent change of station type arrangement. These people served not only in Vietnam but also at the support bases—from Misawa to Guam to the Philippines. During 1964 and 1965, WRAMA had more volunteers than slots; however, as the nature of the war and public opinion changed, the number of volunteers declined. Efforts to continue sending a supply of high quality personnel for these positions remained a primary command concern throughout the war.

Weapon System Support

Of the forty-four aeronautical systems used in Southeast Asia, WRAMA managed twenty-eight. WRAMA had more helicopters, planes, and component systems/sub-systems in Vietnam than any other AMA. With so many systems in use in Southeast Asia, it frequently was said that WRAMA provided more spares than the other four AMAs combined. WRAMA's system and item managers decided whether items should be bought, shipped, stored or modified.

System or Item Management: How's that again?

The classification of system management and item management greatly contributed toward improving the overall performance of system support. By "system," one refers to an aircraft and/or component, technique, operation or groups of components; therefore, system is an ambiguous term. In terms of management, however, we are speaking here of weapons systems.

"Item" applies to the individual parts making up the weapon system. AFLC assigned item management and system management to the AMAs based on current assignments (seeking to balance workload) and past performance. Assignments have been energetically pursued by the AMAs who emphasized factors which favored assignment to their AMA. The assignment of an aircraft to a particular AMA did not mean assignment of all systems on the aircraft.

ASD purchased the system following approval by the Air Force and the using commands. System and item managers worked with the contractors and ASD from the beginning (one of the reasons for the early assignment of the F-15) until the production of the final aircraft, through modifications, and until the final aircraft is scrapped—cradle to grave management.

The "C" series aircraft, C-141, C-130, C-123, C-119, C-54, C-47 and C-7, were utilized in Southeast Asia, for troop and cargo transport, defolient spraying, refueling, gunships, administrative aircraft, and for rescue missions. C-130s and C-141s, managed by WRAMA, served as the long distance cargo haulers of

The HH-53B was the largest, fastest, longest-ranged helicopter used by the Air Force in Southeast Asia. These WRAMA logisticians were viewing the first production model in September 1967.

the war. Flight crews flew C-130 and C-141 aircraft on the "pipeline" (WRAMA to California or Alaska, to Japan, to Vietnam) taking necessary materiel to Southeast Asia.

Twelve WRAMA managed helicopter and utility aircraft systems were active in Vietnam. The 0-1 "Birddog" (a Cessna) was used to lead fighter bombers in on air strikes and as an observation aircraft. Helicopters (used for administrative, rescue and troop carrying) were used by the Navy, Army and Air Force. WRAMA formed agreements with the other service branches for support "in theater" as much as possible. (WRAMA's UH-1F helicopters obtained common parts from the Ton Son Nhut Army Depot.)

WRAMA also assisted other services. For example, in 1965, the U.S. Army's First Cavalry Division and 150 tons of cargo were flown out of Robins by C-130s and C-124s. This operation required round-the-clock duty by WRAMA personnel. The Army, Navy, and Air Force worked together on a number of projects, such as Red Horse, and utilized common equipment items to facilitate logistics support.

Vehicles of the Army's First Calvary Division line up after driving from Ft. Benning, Georgia to Robins for airlift to Vietnam aboard MATS C-124s.

U. S. Army's 1st Cavalry Division vehicles loaded in a C-124 at Robins bound for Vietnam.

Before leaving for Vietnam, WRAMA aircraft went through the maintenance peaking program. This consisted of accomplishing Time Compliance Technical Orders (TCTOs, a one-time change to a Technical Order as a result of modification, new procedure, or one-time inspection). Maintenance also performed periodic maintenance and inspections and applied camouflage paint and corrosion treatment before aircraft left for Southeast Asia.

Maintenance work ranged from handgun repair to C-130 and C-141 IRAN. The C-130s and C-141s underwent complete inspections, then were repaired as required. Repair and overhaul of propellers and accessories, gyro equipment, parachutes, life vests, etc., all occurred at WRAMA.

Robins was the Air Force's only depot repair facility for fighter-bomber electronic "black boxes." The electronic test equipment alone was valued at $20 million in 1966 and personnel in the electronics field at WRAMA expanded to 1,500 that year. Test equipment enabled technicians to perform manual testing by the use of pre-punched computer cards or computer tapes.

This looks like a pile of junk, but WRAMA maintenance employees put six C-7A aircraft back together from these heaps of torn and twisted metal.

The severity of the battle damage is indicated by this wing section.

Clifford Dove (left) of Irwinville and Augustus Nelson of Warner Robins repair shrapnel damage to the fuselage of this C-7A Caribou.

Big Red No. 1—the first of the C-7As repaired from the scraps shown above.

In addition to the IRAN maintenance procedures and Operation Fast Flow (which reduced C-130 overhaul time by 12 days and C-141s by 15 days), WRAMA also repaired battle damaged aircraft.

The Army transferred management responsibility for the C-7As in 1967, and WRAMA's first maintenance work with these aircraft consisted of several severely battle damaged aircraft. RAM (Rapid Area Maintenance) teams returned these aircraft from Southeast Asia to Robins for repair. The aircraft arrived at Robins on railcar flatbeds. WRAMA workers were able to identify portions of the aircraft by use of photographs and careful detective work. WRAMA met several obstacles since Technical Orders were not available and WRAMA personnel had not at that time received hands-on training on the C-7A. Contractor and Army personnel trained WRAMA maintenance people on TDY, the job was completed within the ninety day time period allowed, and the aircraft returned to service.

WRAMA maintenance workers perform inspection and repair on a C-130.

Operation Fast Flow reduced the number of flow days required to process C-130 and C-141 aircraft through Maintenance shops. It also provided Air Force using commands the equivalent of one additional aircraft per year.

The Gunship Programs

The C-47 "Gooney Bird" was rejuvenated during the Vietnam conflict. This thirty-year-old aircraft which had ferried men and supplies over the Burma Hump and through the Berlin Airlift became a ground attack plane. Managed by WRAMA and modified by Air International of Miami, Florida, the first modified C-47 was tested in Vietnam for a year before an entire squadron was brought to Tan Son Nhut Air Base in 1965.

The C-47 was equipped with new communication and navigation equipment, added fuel tanks, a gun sight, and three 7.62 mm Gatling guns (miniguns). The AC-47 was nicknamed "Puff" on its first mission. Firing tracers (normally only every fifth round is a tracer) the South Vietnamese described it as a "dragon." This dragon belched 18,000 rounds per minute and was a fearful sight. The C-47 also functioned as a reconnaissance aircraft for both the United States and South Vietnamese Air Forces. WRAMA continued in the "dragon" business with the modification of the C-130 and C-119 throughout the 1960s and into the early 1970s.

The AC-47 and AC-119 gunships were modified by contractors, but in the case of six C-130s the work was done organically. This joint effort of the Air Force Systems Command and WRAMA produced a substantial savings in both leadtime and dollars. Equipped with more sophisticated equipment, the modification (known as *Pave Spectre*) involved several directorates in the project, with Maintenance doing the modification work.

The Maintenance workload increased to such a point during the war in Vietnam that over 7,000 people were working to keep WRAMA's maintenance operations going (2,500 more than during the Korean Conflict).

This "Gooney Bird" equipped with three 7.62 mm Gatling guns was tested for a year in Vietnam. During that time it was nicknamed "Puff the Magic Dragon."

AC-119 gunships had more sophisticated equipment than the AC-47 and became an important part of the arsenal in Vietnam.

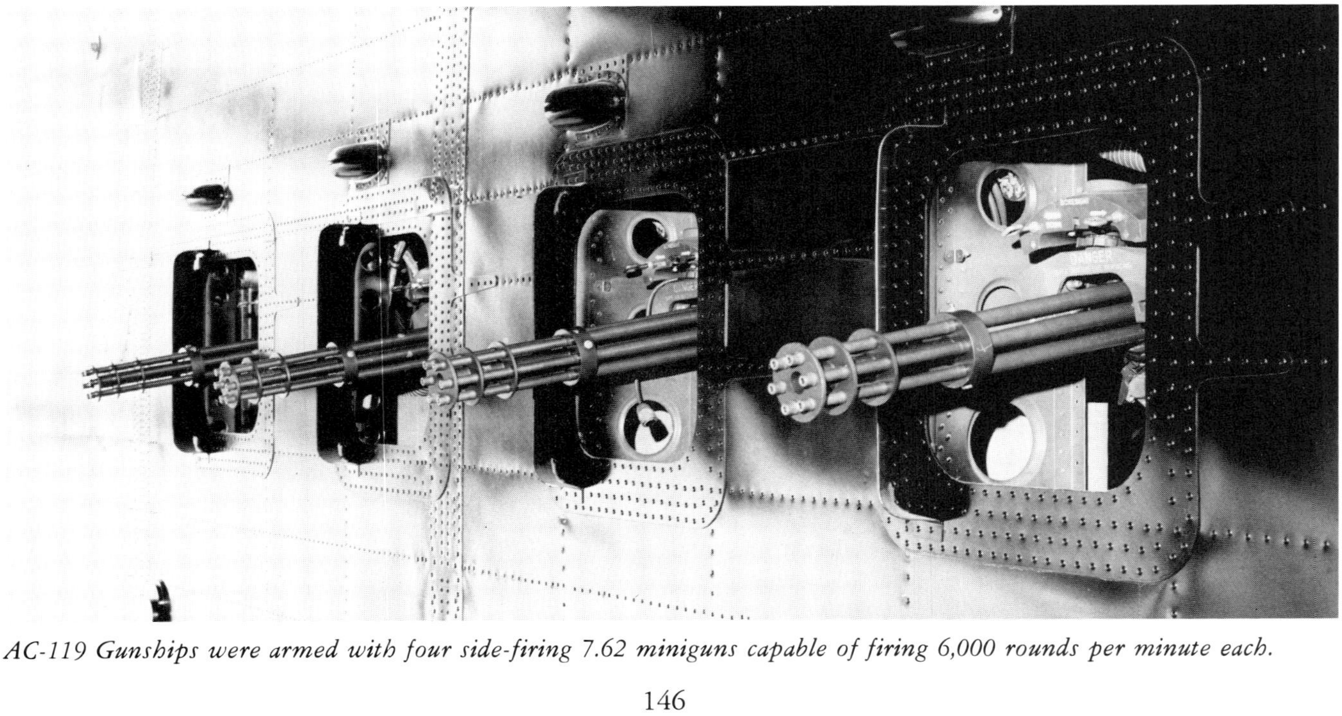

AC-119 Gunships were armed with four side-firing 7.62 miniguns capable of firing 6,000 rounds per minute each.

Electronic Warfare

WRAMA's management of electronic warfare equipment became more important during the war. Electronic warfare was used extensively by both sides (North Vietnam was supplied sophisticated equipment and trained technicians by their allies), and the equipment became more and more complicated. The B-66 aircraft, managed by WRAMA, provided protection for air-to-air combat aircraft, ground air base defenses, countered electronic guidance systems, and was also used to gather electronic intelligence.

Rapid developments in technology and operational requirements required WRAMA to work closely with contractors and the using commands to implement state-of-the-art developments. To do this, WRAMA relocated personnel to the Aeronautical Systems Division (ASD) at Wright-Patterson AFB, Ohio to assist with developing and procuring WRAMA equipment, to contractor plants to expedite delivery of materiel and to the three repair sites for electronic warfare (Oklahoma City, Westinghouse and Hallicrafters). This early venture into electronic warfare support paved the way for WRAMA's assumption of the extensive workloads to come in the 1970s and 1980s.

WRAMA obtained management of tactical missiles from Middletown during phase-up. These missiles, whether radio controlled, laser targeted, guided in on radar emissions, or heat seeking, all received support from WRAMA. Many modifications to make the missiles "smarter" (improved guidance systems) were performed under WRAMA's management. WRAMA also managed the B-57 the BQM-34A drone, the AQM-37A missile target, and the 463L Materiel Handling system.

Repair, routine maintenance, and modifications had to be constantly monitored whether conducted at Robins or at field activities to support the aircraft used in Southeast Asia. Special procedures, such as the Critical Items Office (to isolate and expedite shipping of critical items), the Daily Critical Item Report (provided the item managers with stock levels and asset positions by stock numbers), and obtaining assets from Davis-Monthan storage facilities (where old aircraft were "mothballed"), all enabled item managers to better support WRAMA's systems and aircraft in Southeast Asia.

Procurement utilized teletypes and a follow-up system to ensure that contractors met deadlines. The war resulted in an increase in sole source contracts (ordering only from a specific vendor) due to the increased sophistication of equipment and the critical nature of requirements.

The procedures developed for Project Bitterwine (specific special-purpose ordering, packaging and shipping) contributed to greater efficiency in logistics support. Mechanization improved operations; using systems such as the 463L led to reduced loading/unloading time. Computerized reporting methods allowed item and system managers to keep track of the status of the many systems involved.

Phase-up (1964-1969) and the increased workload created by Vietnam resulted in a shortage of personnel. Training programs, overhires, hiring new personnel, relocating personnel from lower priority workloads, and overtime all helped WRAMA get the job done.

F-104s such as these were equipped with ECM (electrocountermeasure) pods to provide electronic warfare capabilities.

The BQM Target Drone was used for testing air-to-air missiles during the 1960s.

Loading a UH-1 in a C-124, both managed by WRAMA and used in Vietnam.

SAC—on Alert.

Strategic Air Command (SAC)

The Strategic Air Command (SAC) is the primary nuclear delivery force of the Air Force. SAC B-52s all over the world provide a rapid response capability in case of attack. SAC also has the ability to adapt to a conventional warfare role (as in Vietnam). Construction began in the late 1950s in preparation for the arrival of the 4137th Strategic Wing, the first SAC unit at Robins. On March 24, 1959, the unit was activated, and on July 1, 1959, Robins became the wing's permanent home. One officer and one airman arrived in May 1959 to supervise the wing's arrival. The wing received B-52s in 1961.

The 4137th remained at Robins for only four years. On February 1, 1963, the unit was deactivated and replaced by the 465th Bombardment Wing. On July 25, 1968, the 465th was deactivated and the 19th Bombardment Wing came to Robins.

The 19th Bombardment Wing is one of the oldest flying organizations in the Air Force, having been activated on October 18, 1927. In World War II the 19th Bombardment Group operated out of the Philippines and Australia participating in Japan, Papua, Guadalcanal, and Western Okinawa from which its men and machines flew in every campaign. After Korea the 19th Bombardment Wing moved to Pinecastle AFB, Florida, with B-47s and then moved to Homestead AFB, Florida, with B-52s.

The Vietnam conflict had escalated when the wing arrived at Robins in 1968. In June of 1972, the wing began deployments to Guam, the Philippines, and Thailand in Operation *Arc Light*. The wing's tankers participated in Operation *Young Tiger* for the duration of the war. In May 1972, the whole 19th Bombardment Wing deployed virtually overnight to Guam as part of Operation *Bullet Shot*, a program designed to increase U. S. capability to bomb North Vietnam. The wing remained on Guam for nearly eighteen months, participating in the heaviest aerial bombardment campaign in history—*Linebacker*.

In October 1973, the 19th returned to Robins as abruptly as it had left. President Nixon ordered virtually all *Bullet Shot* personnel and aircraft home as the 1973 Yom Kippur War in the Middle East escalated world tensions. By the end of October 1973, all 19th personnel were home.

Since 1973, the 19th Bombardment Wing has been on alert as one of the legs of the U. S. Strategic Triad. During this period of relative peace, the men and women of the 19th have continued to distinguish themselves.

The KC-135 is an important part of SAC's mission. The aircraft provides refueling allowing aircraft to remain airborne. KC-135s also refueled century series aircraft and F-4s during Vietnam.

B-52s provide a rapid response capability.

Detachment 13, 15th Weather Squadron

A weather detachment is located at almost every Air Force base having an operational flying mission. Robins is no exception. Detachment 13, 15th Weather Squadron (Det. 13, 15WS) has been at Robins since 1967. Det. 13 has provided briefing and meteorological watch services to support the critical test flight program of the Warner Robins Air Logictics Center. These services have helped to ensure a successful, safe flying operation.

The mission of Det. 13, 15WS is to provide or arrange for the environmental services needed to support the exercise, contingency, and wartime requirements of the ALC and the 19th Bombardment Wing; to provide staff weather officer support to HQ AFRES; and to provide routine environmental services to all Robins AFB units and transient aircrews.

In addition to the daily routines, the squadron is also involved in the many exercises and war games conducted each year. Approximately thirty-seven percent of all squadron personnel are on mobility status. The recent super exercise *Bright Star 82* (the deployment/employment and redeployment of the Rapid Deployment Force to the Middle East) was an excellent example of many people in the squadron working together, both stateside and overseas, to help make the deployment and redeployment a success. Fifteenth Weather Squadron personnel were the first weather folks in on *Bright Star* and the last out.

In October 1981, Det. 13 obtained a capability to receive weather satellite information. Satellite data is an invaluable forecasting and briefing tool that will improve weather services at Robins.

Det. 13 provided special briefing and intensified meteorological watch services to support distinguished visitors meeting with former President Carter at Plains, Georgia. This high visibility support was provided with professionalism and enthusiasm.

TSgt Sheldon Meyers and Sgt. William Gladden inspect an AN/GMQ-13 Cloud Height Set.

Sign at entrance of Robins AFB SAC area.

On July 8, 1979, SAC conducted its first command-wide exercise in twenty years. SAC units world wide responded to an exercise scenario which simulated escalating world tension by gradually increasing their states of readiness. At the peak of the exercise, one hundred B-52s from the United States and Guam were launched on simulated war missions to demonstrate the scale of SAC's response to attack. Bombers and tankers from the 19th flew missions ranging from simulated strikes to shows of force, obtaining exceptional results.

In 1981, the 19th Bombardment Wing established a standard which may never again be equalled. First, during *Global Shield* 81 its bombers achieved the third best results of any B-52G or H unit. In September, members of the 19th Munitions Maintenance, 19th Organizational Maintenance, and 19th Security Police Squadrons competed in the SAC munitions loading competition, *Giant Sword.* These expert airmen won first place, bringing home the coveted Stanley A. Barrentine Trophy.

September also saw the wing's bombers achieve the best score in command history on a flying evaluation, earning the prestigious General George C. Kenney Trophy. In 1981, the 19th Bombardment Wing Maintenance squadrons were named "Best in SAC," and the wing maintenance operation named one of three finalists for the Daedalian Trophy, which recognized the best maintenance in the Air Force. Also during 1981, the tankers of the 912th Air Refueling Squadron had a rare one hundred percent on-time launch record for higher headquarters directed support sorties which ran the spectrum of SAC refueling missions. Finally, Captain Phillip D. Booth of the 28th Bombardment Squadron received the General Bruce K. Holloway Humanitarian Award citing him for the most outstanding community service in the command.

The 19th Bombardment Wing hosts two of the finest tactical flying squadrons in the command. The 28th Bombardment Squadron is the oldest flying unit in the United States Air Force, and has achieved fame in four wars. The 912th Air Refueling Squadron has provided reliable aerial refueling support to every possible receiver for nearly three decades.

The Black Knight, the symbol of the men and women of the 19th Bombardment Wing, has been passed through the three units at Robins. It was originally awarded to the 4137th Strategic Wing in a 1961 Combat Crew Competition. As the 4137th and, later, the 465th Bombardment Wing were deactivated, the suit of black armor, and its legacy were passed on. Like the chess knight, able to move in any direction, he symbolizes the 19th Bombardment Wing's maneuverability, global range, and tactical posture. These attributes are summed up in the unofficial wing motto, "Checkmate to Aggression."

Construction

The addition of new assignments resulted in increased construction during the 1960s. When the first SAC aircraft arrived in August 1960, the majority of SAC-related construction was completed. These new facilities, begun in 1958, consisted of runway extension and enlargement which gave WRAMA a 12,500 foot long runway, new taxiways, 15-20 inch thick concrete aircraft aprons, an instrument landing system transmitter, a squadron operations building, twenty-four jet blast deflector fences, jet fuel hydrant system, operational maintenance docks, POL (petroleum, oil, lubricants) facility, and additional housing. A two-story readiness crew building having a seventy-man capacity and containing sleeping quarters, kitchen, dining area, and an alert ramp leading to the runway was built.

Military housing increased with the construction of 423 units of Capehart on-base housing (2 units were later destroyed by fire), begun in 1959. The unsightly Zeigler Apartments were demolished and the land sold to the city. Three hundred additional units now known as Hillside were built between 1964-1966. Two two-hundred man airman dormitories and a forty-man officers living facility were constructed for unmarried military personnel. The seventy-five-lot trailer park, built in 1965-1966 in the "Dormitory Area" gave residents paved walkways, large yards, and adequate lighting. To aid personnel in finding off-base housing, a Housing Referral Office was established in Building 274 to give rental guidance to WRAMA's servicemen.

In the intricate work Betty Collins was performing as a gyro repairer in a "clean room," cleanliness was important to insure perfect operation of equipment.

By 1962, Maintenance shops consisted of two and a half million square feet of floorspace with an additional four million square feet of concrete hardstand for parking and working on aircraft. Advances in technology required modernization of plant facilities. Systems having space age technology and needing modern shops were being worked on in World War II facilities. Building commenced in 1960-1961 to convert two old buildings into a clean room facility for gyro repair and a chemical laboratory.

Problems with corrosion had led to WRAMA being named Corrosion Control Center for all the Air Force. A facility was needed for testing and painting. The new aircraft paint facility, built in 1965 at a cost of $991,176 gave WRAMA the capability to provide aircraft paint and corrosion treatment. The building could accommodate two C-130s or one C-141. During the 1960s, WRAMA made a concerted effort to perform more maintenance work in-house. The Gatling gun test facility allowed Robins to perform testing following overhaul. The Army had previously done these tests.

The "laminar flow" booth at the rear provided "clean room" conditions by removing dust particles from the air. These employees were working on B-52 bomb-navigation systems.

New Paint Shop built during the 1960s, a part of WRAMA's industrial modernization building program, can serve one C-141 or two C-130s.

Increased use of electronics required improvements to shop areas. Sensitive equipment had to have controlled temperature and humidity to operate properly. Many of the newer systems purchased during the 1960s required temperatures of no more than 72 degrees, humidity control levels of forty percent, and dust controls. Georgia's hot climate with humidity levels sometimes hovering near ninety percent were not conducive to specialized repair on sensitive electronics without clean room conditions.

The acquisition of the C-141 resulted in additions to docking areas and a new aviation and hydrant refueling system which permitted rapid fueling/defueling of aircraft. "Operation Fast Flow" resulted in the purchase of $860,000 worth of workstands and the installation of bridge cranes to allow more accessibility to various parts of the aircraft at one time. Industrial plant modernization was needed to provide R&D (Research and Development) and testing in support of system engineering. WRAMA's R&D consisted solely of projects related to redesign and reengineering of WRAMA managed systems. As a part of this effort, Building 155 was converted to a technical laboratory to test electronic test equipment and components. The new electroplating and metal finishing facilities were also part of the plant modernization.

Increased workloads from phase-up and Vietnam created parking and traffic problems. Several parking lots were restriped to crowd the cars and allow more parking. Carpools were also encouraged. The construction of acceleration/deceleration lanes on 1.7 miles of Highway 247 allowed easier access onto and off of the base. WRAMA and the city worked together to get railroad warning signals for the crossings of Elberta Road and Davis Drive where a number of tragic deaths had occurred over the years. Efforts began during the 1960s to obtain a direct connector linking the base to I-75.

Gatling Gun undergoing repair. In the 1960s WRAMA constructed a test facility for these weapons.

Telephone communications improved with the addition of autovon lines.

Communications improved in 1959 when Southern Bell installed an inward/outward dial system for the base. In 1963, WATS (wide area telephone service) lines were extended to cover the entire state of Georgia. The greatest improvement came with the addition of AUTOVON lines. AUTOVON consists of a telephone network which links defense installations all over the world. (Today, these lines are inadequate for the constant usage demanded and new lines are being added.) The new base communications center provided 12,700 square feet for cryptographic equipment to process incoming and outgoing messages. In 1964, the WRAMA Command Post received a KY-9 phone scrambler on which top secret information could be transmitted.

WRAMA's computer center was the largest in the Southeast when it opened in 1964. By 1975 (when this photo was taken) data processing systems affected virtually every job on base.

In 1964, the opening of the Computer Center, the largest automatic data processing center in the Southeast, ushered in the "Age of Computers." Computers were used for payroll (WRAMA processed Dobbins' and Moody's payrolls in addition to the Robins' payroll), inventory listings for all Air Force vehicles and various listings of WRAMA prime items. Employees received special training in computer languages as computerization was extended into almost every job on base. By the 1980s the impact of computers on the work of all employees was nearly complete.

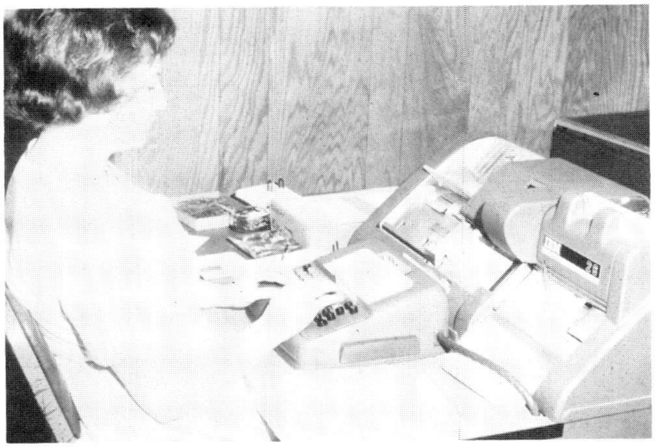

Keypunch operator jobs increased with the increase in computerized data systems.

Recreation facilities improved with the construction of the sixteen-lane bowling center, a new gym, lighting for tennis courts, camping and scouting facilities, picnic areas, and an addition to the base chapel. The NCO Club obtained a swimming pool through a "do-it-yourself" method. The funds were obtained from the AFLC Loan Fund and paid back from an increase in dues. The pool was dedicated in August 1969.

Other construction projects included the conversion of warehouse areas of Building 300 into office space due to increased need for office space resulting from phase-up and Vietnam, the air police building, the jet fuel pipeline (from Texas to WRAMA) was built to reduce the high cost of transporting fuel, an industrial waste treatment plant, two elementary schools, a sixty-bed hospital, and renovation to existing buildings.

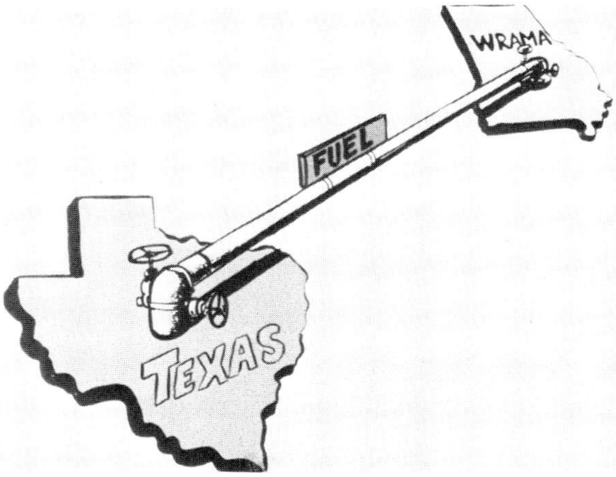

Pipeline connector to carry fuel directly into the base was built in the 1960s.

These Zeigler Apartments were razed so that modern housing could be built for WRAMA serviceman.

Housing

In 1962, an Air Force official testifying before the U. S. House Appropriations Sub-Committee said, "The housing units occupied by lower ranking airmen

Three hundred units of Capehart housing were built in the late 1950s and early 1960s.

(at Robins AFB). . . are probably the worst I have ever seen in the Air Force . . . approximately slave quarters." Housing presented problems for Robins off and on until the middle to late 1960s, depending on workload.

In the early 1960s a major push for adequate low-cost rental housing grasped the attention of local newspapers and politicians. There were discussions in local newspapers of the impact on the economy if airmen had lower rents (They would have more money to spend!); of how improved housing could attract other industries; how low morale would result from airmen separated from their families due to lack of housing; and how WRAMA might lose workloads if housing conditions did not improve.

SAC, in an action unrelated to housing, temporarily went to Westover, Massachusetts, for a time during 1962 while runways were resurfaced, emphasizing the fact that workloads could be moved. This impressed local officials more than anything else. This move, even though of a temporary nature, pointed out the economic impact if even a small part of the WRAMA contingent were reassigned.

Work began by Middle Georgia leaders to obtain Public Housing Administration low-cost housing for lower income families. The tearing down of substandard Zeigler housing allowed replacement with newer housing for military personnel. Prior to the construction of these new units, airmen had lived in Zeigler apartments, trailers, or whatever housing was available in the area. Travel time, distances from base, and availability of housing limited choices for both military and lower grade civilians.

By 1964-65, with the beginning of phase-up and Vietnam workload increases, large numbers of private dwellings and military housing had been built. Local officials emphasized the recent housing built, the age of houses, and the variety of housing available to potential functional transferees and new personnel coming into the area.

By the end of the 1960s, the housing situation no longer seemed impossible. Waiting lists for base quarters still existed, many airmen not able to afford purchasing a home resided in trailers and rents frequently exceeded housing allowances, but the quality of housing improved. More living space, cleaner conditions, a new military trailer park, and larger yards all contributed to better living conditions.

Civil Rights

Black Americans, by the 1960s, were unwilling to settle for an unequal role in American life. Such actions as sit-ins, protest marches, voter registration drives, and desegregation of previously all-white establishments met with success in some areas and with violence in others. Vast changes would come with the growing social consciousness of the 1960s.

Highly controversial topics such as civil rights, elimination of poverty and women's issues reflected a new awareness of individual rights. The concensus was that America's efforts needed to be channelled toward the development of programs that would effectively help those who could not help themselves.

The Armed Services, following President Truman's order to integrate the services in 1948, represented one of the few areas where blacks and whites lived and worked in equality. By working and living in closer contact, black and white servicemen dissolved barriers which had previously kept them apart and both benefited from the experience.

From the beginning, Robins attracted people for the higher salaries and job opportunities not available elsewhere. The agrarian economy of Middle Georgia had ill prepared either group for the technical work required at the depot. Blacks were concentrated in occupations with little or no future i.e., laborers, loaders, janitors, etc.

This display reflected early "Affirmative Actions" supporting Equal Employment Opportunity.

The Civil Rights Act of 1964 provided for equal employment opportunity regardless of race, color, religion or national origin. At WRAMA, committees were established to formulate an action plan to implement Civil Service guidelines to eliminate discrimina-

Thomas L. "Tom" Delton

General Gideon and Thomas L. Delton discuss the Equal Employment Opportunity Program for Robins.

Tom Delton, Deputy Equal Employment Opportunity officer, was always a firm believer in education. From his home in Mansfield, Louisiana, he traveled on a freight car with $1.50 to go to Tuskegee Institute. After graduating in 1937, he moved to Toombsboro, Georgia, to teach agriculture. There he began the first NAACP chapter in that area. He served in the Pacific during World War II with the 372nd Infantry Regiment. He returned to Georgia following his military service and continued working with the NAACP, participating in a number of civil rights marches in the 1950s and 1960s. In the early 1960s, Mr. Delton worked for the United States Department of Agriculture as a Rural Community Development Specialist. This job brought him in contact with low income farmers to whom Mr. Delton introduced the various federal farm programs available to them. One part of his job was establishing food programs for Head Start operations thoughout the Southeast.

Mr. Delton came to work at Robins in 1967. His extensive work in civil rights initiatives was a definite asset in the EEO program to end discrimination at Robins. His ability to get both sides to sit down and talk together proved invaluable. He felt that if blacks and whites would communicate and learn more about each other, their attitudes might change. Education and motivation were critical factors in upgrading black workers. Tom Delton frequently stated that EEO was not there to give "hand outs" but that qualified people had to be found. He encouraged blacks to take advantage of educational and training opportunities to obtain the skills necessary for advancement.

Mr. Delton remained in the Army Reserves until 1972, retiring as a Lieutenant Colonel. He was instrumental in locating a $400,000 Armory Training Center in nearby Fort Valley. He retired from his position at Robins in January 1980.

tion at Robins Air Force Base. Civilian personnel conducted meetings with supervisors and employees to explain the plan and outline requirements. Supervisory development courses were revised to include information on EEO in an attempt to change attitudes and behavior. New employees received information on WRAMA's EEO programs during initial orientation briefings.

WRAMA stressed hiring of non-skilled persons for entry level (GS 1-2) positions to employment agencies, high schools and colleges. Visits by WRAMA recruiters to minority colleges obtained graduates for GS-5 positions. The EEO program worked to motivate minority employees to participate in educational programs. WRAMA encouraged the development of technical education at Fort Valley State College (historically a black college) to provide entry-level competence into underrepresented technical areas.

Project Youth provided the young people a variety of recreational opportunities.

Asst. Secretary of Defense, Roger T. Kelly, observing Robins AFB summer youth program handicrafts.

Project Youth

Robins AFB provided a unique opportunity for children aged ten to twelve from eight Middle Georgia counties in the Summer Youth Program. Various educational and recreational skills were taught to the youngsters. Tours of the base, horseback riding, bowling, golf and swimming formed a part of the five-day program. The young people represented an economic and cultural cross section of their communities. Transported by bus, many came from as far away as Bleckley and Dodge Counties. The Air Force Association provided free hot lunches (food was obtained from the United States Department of Agriculture). Many of these children later became base employees.

Project Youth participants came from counties in the surrounding area.

The Civil Rights Act of 1964 prohibited discrimination in business, housing and recreation. Although not a part of the WRAMA mission, programs launched in the middle and late 1960s to obtain equal housing met with some success. Surveys conducted of local realtors by the base provided listings of housing available to both blacks and whites.

Eight years after the famous Brown vs Board of Education decision rendered "separate but equal" schooling illegal, on March 30, 1962, Abraham Ribicoff, Secretary of Health, Education and Welfare, announced to the House Education Subcommittee that beginning in September 1963 segregated schools in communities having children living on military bases would not be suitable and that such schools would lose their federal funding if they remained segregated. He also stated that schools could and would be built on military installations if local school boards refused to desegregate.

Helen Layfield operating a heavy duty sander in the 1960s.

The decision was made by HEW and the Department of Defense to end part of the problem by building an elementary school on base, taking the pressure off the county for a short time. Robins Air Force Base Elementary School opened in September 1963. By 1965, the base school population had increased to such a point that a second school, Linwood Elementary, was built.

Working at Robins Air Force Base had reduced many of the prejudices held by whites. By working with blacks on an equal basis, many whites accepted integration more easily than their counterparts in other areas of the South. The number of military and retired military personnel and their dependents who had lived with integration for years also eased the process. The presence of the base did more to ease the process of integration in Warner Robins than any other factor.

By the end of the 1970s, the adoption of Affirmative Action plans produced specific goals to be met. By 1980, the focus of EEO had split into complaint processing and the original goals of equal hiring, promotion and treatment for all employees. Blacks, women, Hispanics, Indians and Orientals worked in numerous types of jobs in which minorities had been underrepresented ten years earlier. Educational opportunities, improved communication, and fair employment practices had virtually eliminated discriminatory practices.

During World War II, out of 1800 people in the Engine Repair Shop, 62 percent were women.

With the men off at war, women emerged from the kitchen to work as bricklayers, aircraft mechanics, parachute packers and at hundreds of "men's jobs" to support the war effort. In 1945, out of eighteen hundred people in the Engine Repair Shop, sixty-two percent were women. When the war ended and layoffs began at WRAMA, many women sought alternative employment. "Rosie the Riveter" discovered she preferred working, liked accomplishing any task set before her and valued the self-esteem gained from a job well done. Jobs in secretarial fields remained an opening for the woman who wished to work. Many women used the office to start to work their way to the top as in the case of Robins' two highest ranking women.

Eleanor Richardson, WRAMA's first woman GS-15, was at one time the highest ranking woman in the Air Force.

A native of Quitman, Georgia, Mrs. Richardson's father owned a general store which dealt more in barter than money. She graduated from Quitman High School in 1933 and began work with the Department of Agriculture the following year. In 1942, she began working in the personnel department at Moody Field, Valdosta, Georgia, for the Flying Training Command. In 1946, Mrs. Richardson came to Robins as Chief of Classification and Wage and in 1956 became Chief of Management Services, which later became the Plans and Management Office.

In 1963, Mrs. Richardson competed against forty-one other GS-14s, nominated Air Force wide, for the new position as Deputy Director (GS-15) of Plans and Management. On being selected, Mrs. Richardson became the Air Force's highest ranking civilian woman and the only woman Deputy Director at any of the AMAs.

Mrs. Richardson was married to Roy E. Richardson, Chief of the Base Procurement Division of the WRAMA Procurement and Production Directorate. The couple had one son, Michael. Mrs. Richardson retired in December 1969; she died in 1982, following her husband's death the previous winter.

Vivian Hartman came to Robins from Screven, Georgia, at age sixteen to work as a clerk-stenographer in Procurement. From Procurement, she went to Statistical Services to work on punchcard accounting machines, the forerunners of today's computers. As a natural progression, Mrs. Hartman moved into computer programming, working with the early (first generation) computers. In 1964 she began eleven years in Plans and Programs as a program manager for projects relating to computerization. In 1975, Mrs. Hartman returned to Accounting and Finance and rose from Branch Chief to Division Chief to her current position as Deputy Comptroller—the second woman at Robins to achieve a GS-15 position.

In the course of her career, Mrs. Hartman obtained a degree in Business Administration, "took all the dirty jobs no one else wanted," long hours, TDY to obscure places, shift work, and "hard work and more hard work."

Being the highest ranking woman on base, today, Mrs. Hartman feels that she must serve as a role model for other women and her duty is to help other women with their careers. In achieving success, she encourages women to get ahead, to obtain education and training, to participate in the Federal Women's Program and to take advantage of all opportunities which come along.

Mrs. Hartman and her husband, Harry, have three daughters Karen, Sheree, and Vivian Lee.

Eleanor Richardson (left) and Vivian Hartman (right) confer with Herbert Hibler (OCAMA), Harry Stiles (AFLC), and General Gideon on the Selective Overhaul Program in 1966.

Community Relations

Over the years, WRAMA employees have expressed their concern for their fellow man by participation in charity donations (such as Combined Federal Campaign), blood drives, and through membership in various civic organizations. Commanders and high ranking civil service employees have served as speakers time and again on a variety of topics to high schools and service clubs. WRAMA received praise from both AFLC and the Air Force throughout the decade for community involvement and for the willingness of its people to voluntarily participate in projects to improve life in Middle Georgia.

One of the best examples is Macon Spring Clean '69. The tempo of the times played a large part in the success of the Spring Clean. The willingness to donate time and energy into volunteer projects to help others was all part of the altruism of the period. The "ask not what your country can do for you" inaugural speech of President Kennedy set an atmosphere in which people wanted to make their country better.

Painting and construction repairs were part of the effort.

In 1969, three thousand WRAMA employees volunteered for Macon Spring Clean '69 and transformed a seventy block area in the Tybee section of Macon into a model city. The spring cleaning idea, originated by the Citizens and Southern Bank, had been tried in Savannah and helped earn that city an All-American City Award. Macon's C & S Banks felt

Robins people gave freely of their time during Macon Spring Clean '69.

These Robins employees provide structural repair to an old house.

that the project was worthwhile and enlisted the assistance of local governments, schools, and Robins Air Force Base. Volunteers hauled away over two thousand tons of trash, removed three hundred junked cars, cleaned yards, painted buildings, and provided construction repairs. Secretaries, executives, officers, enlisted men, blacks, whites, strangers and friends all worked side-by-side to upgrade the living environment of the area.

Letters of appreciation for WRAMA's efforts from Macon business and civic leaders reached the Secretary of the Air Force, and WRAMA in turn was praised for its community involvement.

In 1970, the Pleasant Hill area of Macon was the target for the clean-up operation. WRAMA's Civil Engineers created a new park to provide recreation for the residents. The City of Macon expressed its appreciation to WRAMA by naming the new park for the WRAMA Commander, General Beck—the first facility in Macon named for a WRAMA officer.

Following Macon's example, Warner Robins held Spring Cleaning in 1970. The entire city received attention as the volunteers removed abandoned vehicles, refrigerators, washing machines, trash and debris.

WRAMA's citizenship was state and nation-wide. WRAMA furnished aid to the victims of the Alaskan Earthquake, Hurricane Camille, tornadoes in Florida, etc. A crippling ice storm hit Middle Georgia in 1964 and the base assisted local communities with emergency generators, blankets, food, and transportation. During the ice storm, a number of dairy farms lost electricity. Their cows, used to electric milkers, would not allow manual milking. WRAMA saved the day (or milk, in this instance) by lending portable generators to the farmers—just doing its part as a good neighbor.

Aerial view of Spring Clean target area, Tybee section of Macon.

Robins' Charming Ambassadors

The first contestant to be crowned "Miss Robins Air Force Base" was Ann Walters, a secretary in the Directorate of Supply and Transportation, who won the title in September 1966. A blond, nineteen year old native of Lumberton, North Carolina, she represented the 19th Field Maintenance Squadron of the Strategic Air Command in that first pageant. Today, Ann Walters Oliver is vice-president of the Houston Federal Savings and Loan Association in Warner Robins. She is still as beautiful as ever.

The current "Ms. Robins Air Force Base" (to be unmarried is no longer a qualification) is Patty Brittian Burkett, a secretary in the Directorate of Transportation of the Air Force Reserve Headquarters, who won the title in November 1981. The twenty-five year old brunette was born at Fort Riley, Kansas. She represented Hq AFRES in the 1981 pageant.

The purpose of the Robins AFB pageant is to select a base representative to greet visiting dignitaries, attend various awards ceremonies, etc.—to add charm and beauty to the occasions.

Ann Walters, the first Miss Robins AFB, September 1966

Patty Burkett, the current Ms. Robins AFB, with Georgia's junior senator, Mack Mattingly, 1982.

The Carl Vinson Memorial Chapter 296 of the Air Force Association

The Air Force Association, founded in 1946 by members of the World War II Army Air Forces, is an independent, non-profit organization dedicated to obtaining and maintaining adequate aerospace power for national security and world peace.

The drive to organize a local group began in April 1963. Lt. Col. Alan E. Goldsmith, Information Office, became the focal point for the membership campaign and had the hearty indorsement and encouragement of General Hefley, WRAMA Commander. At the Sandman (Civilian Club) on July 25, 1963, the charter for the Middle Georgia Squadron was signed and the first slate of officers were elected. They were: Squadron Commander: Jack A. McMurtrey, FAA; Vice Commander: James Carey, Physician; Secretary: Jeff Pierce, First Federal Savings & Loan; and Treasurer: Stephen J. Chasse, WRAMA Comptroller. John Crandall and Joe Childs of Macon and Tom Lindsey and Wallace Reid of Warner Robins were elected as council members.

The week of 6 to 12 October 1963 was proclaimed "AFA Week" by Governor Carl Sanders. A week-long schedule of activities culminated with a banquet at the base restaurant on October 12. At that time, the charter was presented. The squadron was made up of seven hundred members (two hundred AFA members who changed their chapter affiliations to Middle Georgia and five hundred new members).

During the remainder of the 1960s, local organization leaders included J. D. "Dixie" Walker, Edward Barker, Paul K. Fain, and Dr. Dan Callahan. Dr. Callahan, the only chapter president to serve two terms and become state president,[8] had a very active administration. During his terms as president the MACE Missile was erected near the Warner Robins exit on Interstate Highway 75, the Air Force Junior ROTC was introduced into the local high schools, a state-wide campaign for information and release of the Prisoners of War (POW) and Missing in Actions (MIAs) was mounted, and a scholarship program for the Robins Resident Center was established. Dr. Callahan coined the slogan, "Every Day in Middle Georgia is Air Force Appreciation Day (EDIMGIAFAD)."

William "Bill" Wisse, William "Bill" Powell, Ken Greer, Don Allen, H. C. "Butch" Strawser, C. E. "Pat" Mahaffey, Betty Clark and L. A. "Buster" McConnell chaired the chapter during the 1970s. The chapter membership grew to over 2,000 and the number of life members to over 150.

During the 1980s, the chapter presidents have been Jack Maret and Bobby Bates, with Wilbur Keck serving as the current president.

The chapter has donated over $42 000 to various base and community Air Force related projects since 1970, including $3,500 in educational grants, $1,500 to the Air Force Museum Foundation, $4,700 to the Summer Youth Program and $3,500 to build a picnic shelter on base. The chapter has also sponsored the AFRES Band Concerts since 1973 with the *Daily Sun* as a co-sponsor in recent years.

The Carl Vinson Memorial Chapter today is over 3,000 strong, has 178 life members, and is one of the most decorated chapters in the nation. It remains a vital part of both the base and community life.

Congressman Jack Brinkley (Third District, Georgia) addressed a meeting of Middle Georgia Chapter 296 of the Air Force Association in 1969. Joining Mr. Brinkley are J. D. "Dixie" Walker (Georgia State AFA Vice President), Brig. Gen. Ralph Holland (then WRAMA Vice Commander), and Dr. Dan Callahan, (President of the local Chapter, later AFA National Vice President).

[8]Other local members who became state president included Mr. H. L. Everett, Gen. James D. Thurmond and Mr. Lee C. Lingelbach.

C-141

The discussions of a cargo aircraft built solely for military use began during World War II when so much cargo was ferried over the "Hump" to Burma. Later lessons in the logistic supply of the Berlin Airlift and during the Korean War showed the need for "large aircraft, both for inter-theater and intra-theater airlift."

Military Air Transport Service (MATS) recognized the limitations of the C-130 in the "flexible response" foreign policy prevalent during the late 1950s and early 1960s. The Eisenhower Administration first supported building the C-141; Democrats and Republicans were in agreement. At a speech to the American Legion in Miami, Florida, President Kennedy stated that it was useless to have seventeen divisions if they could not be transported anywhere in the world in twenty-four or forty-eight hours. During the congressional hearings for the C-141, MATS performance during operational exercises conducted in the Carribbean convinced many congressmen that the new aircraft was needed. Lt. Gen. William H. Tanner, MATS Commander, wanted "a work horse plane, cheap to operate, of indifferent speed, relatively large and easy to load."

Comparison of C-141 and C-130.

WRAMA officials were present at the C-141 Rollout Ceremony August 22, 1963.

General Hedlund prepares to take the C-141 up for an aerial tour of Middle Georgia.

C-141 with workstand in place for maintenance work.

Boeing, Douglas, Convair and Lockheed were competing for the contract. Among the design requirements was the use of 463L loading system, which was managed by WRAMA. On February 8, 1961, WRAMA was named system manager of the C-141. President Kennedy announced that Lockheed had been awarded the contract on March 13, 1961.

On August 22, 1963, Lockheed rolled out the first C-141 with Air Force officials (including WRAMA's commander, General Hedlund) in attendance. The first flight was made on December 17, 1963, and the first C-141 delivered to MATS at Tinker AFB, Oklahoma on October 19, 1964. When the first C-141 at WRAMA arrived on July 20, 1964, General Hedlund took the aircraft up for a local flight over Middle Georgia.

By 1965 the C-141 was providing Military Airlift Command (MAC) (The name of the Military Air Transport Service had been changed to Military Airlift Command) the support which had been forecast in the war in Southeast Asia. The C-141 became the backbone of MAC forces. Loaded with cargo going to Southeast Asia, the C-141 could be used for aeromedical evacuation on return flights. The rollers and restraining rails for the pallets could be retracted to provide a smooth surface for bulk loading, rolling stock, troops, and litter pallets. To reconfigure from 463L cargo to any of the other 19 possible configurations required only a few hours (only the VIP passenger configuration required 24 hours). These facts in combination with reduced time for loading/unloading gave the C-141 a quicker turn-around capability.

The C-141 has the capability to span any ocean nonstop at more than 500 miles per hour, airlifting up to 70,000 pounds of cargo or 154 combat troops. With four Pratt & Whitney engines, the C-141 has a maximum cruise speed of 488 knots but can also make aerial delivery drops at speeds of 130-150 knots. The original C-141 had a length of 143 feet, a height of 39 feet and a wing span of 160 feet. The wing span is longer than the entire flight of the Wright Brothers at Kitty Hawk. In the original C-141, the cargo compartment was 70 feet long, 9 feet high and 10 1/4 feet wide.

C-141 in flight.

General Hedlund, WRAMA Commander, discusses WRAMA's worldwide logistics support of the C-141 Starlifter with Col. Donald G. Bussey, chief of WRAMA's C-141 Systems Support Management Division. The two pilots have participated in test flights in the new fanjet from Robins to Edwards AFB, California.

Workloads for the Future—The F-15

The introduction of the system manager concept in 1968 went a long way toward solving problems and needs of operating commands and in improving management concepts in control of logistics support. Each airplane, helicopter, missile, and system was placed under the responsibility of one person who would be in total charge of the satisfactory operation of each system. The system manager integrated and coordinated all actions that were related to a system. More emphasis was placed on user's requests and on communicating reasons for delays to the user.

Another factor of concern at WRAMA was the lack of new assignments. Other than the C-141 and systems acquired during phase-up, WRAMA received few new assignments in the 1960s. General Hedlund had discontinued calling WRAMA the "Cargo Transport AMA" as he felt WRAMA was creating an image which would exclude other workloads. The new fighter-bombers, F-4 and F-111, had not been assigned to WRAMA and even the C-5A went to San Antonio. When discussion concerning the F-15 assignment began, General Beck made a concerted effort to gain this assignment for WRAMA.

System management had provisionally been assigned to San Antonio. General Beck assured AFLC that WRAMA had the managerial capability to handle the F-15. He further stated that WRAMA could accomplish the workload without an increase in either manpower or facilities. WRAMA's reputation had improved substantially by 1969, customer complaints had declined and industrial modernization increased through new building programs. AFLC obviously agreed.

In March 1969 AFLC assigned system management for the F-15 to WRAMA. This assignment, coming before the awarding of the contract, involved WRAMA in the source selection. Fairchild Hiller, McDonnell Aircraft (MCAIR), and North American Aviation were competing for the contract, which had an estimated value of $7.3 billion.

A cadre of thirteen production, engineering, operations, requirements, and item management people, personally selected by Mr. Cotton, began determining the type of data required to make the selection. The Air Force "Request for Proposal," data acquired from San Antonio and from the contractor, and using command requirements were reviewed to establish standards for evaluation of each contractor's proposal. The review team, consisting of approximately two hundred people from TAC, ATC, AFLC, WRAMA, ASD, and AFSC worked at ASD for three to five months evaluating the contractors' submittals.

WRAMA evaluated each contractor's proposal for support equipment, test equipment, technical data, reliability, supportability and other factors bearing on logistic support. WRAMA's portion of the evaluation (twenty percent) indeed turned out to be a critical portion of the selection process. Each contractor received evaluations based on the pre-established standards on a scale ranging from one to ten. This was the first selection board of this kind in which all affected commands participated in contractor selection. McDonnell Aircraft, a division of McDonnell Douglas, was awarded the contract on December 23, 1969.

The competition for Specialized Repair Activity (SRA) was between Warner Robins, Sacramento, and Oklahoma City. For the first time, WRAMA competed for assignment for repair of a fighter aircraft. A brochure prepared by systems manager (SM) and maintenance personnel emphasized WRAMA's assets and the abilities of its workers. The brochure pointed out the cost savings of having system management and specialized repair located in one place: quicker response to users, improved technical competence by having SM/SRA in one place, and maximum response at minimum cost. WRAMA's extensive training program for maintenance workers resulted in a $240,000 savings (from the contractor's requirements). The tools and test equipment, valued at $100,000, also furnished evidence of WRAMA's modern industrial facilities.

Crowd gathered around WRAMA's Eagle.

This F-15 from Eglin AFB provided Robins its first view of the Eagle.

Congressman Jack Brinkley looks at the Eagle.

The greatest asset remained WRAMA's workers. With an average of 16 years overall experience; 280 man-years of fighter experience gained over the past 5 years (1964-1969) on F-4 and F-106 aircraft; and the success of Operation Fast Flow, WRAMA demonstrated that of all the Air Materiel Areas, Warner Robins was best suited for the job.

On July 27, 1972, the first F-15 flew. Between January 16 and February 1, 1975, the nineteenth production F-15 broke eight world altitude records in Operation Streak Eagle. During Streak Eagle the F-15 reached an altitude of 98,425 feet (18.64 miles) in 3.46 minutes.

WRAMA's first view of the F-15 came in November 1974, when an F-15 from Eglin AFB, Florida, visited Robins for a two day Open House celebration. On May 30, 1975, the Eagle offically came to WRAMA as a prototype for engineers and technicians to disassemble, study, and reassemble. This allowed WRAMA to test its procedures, tools, and test equipment in addition to giving workers hands-on experience with the F-15.

The appearance of the supersonic (Mach 2.5 class) F-15 is characterized by a fixed, shoulder-mounted, swept back wing of low aspect ratio and twin vertical stabilizers. Powered by two dual-spool augmented F-100 turbofan engines, its high thrust-to-weight ratio allowed it to break the sound barrier in a straight vertical climb. The engines' air-induction system consists of variable position inlet ramps to provide optimum airflow to the engines, insuring efficient operation in a wide range of aircraft speeds and altitudes.

The low wing loading and high thrust-to-weight ratio of the aircraft are designed to provide exceptional maneuverability for aerial combat, with an advanced electronic system helping to sort and identify targets, and the Tactical Electronic Warfare System (TEWS) enabling it to evade enemy defenses. Although the F-15 is armed with an advanced fire control system with radar-guided and heat-seeking air-to-air missiles, and an internally mounted 20 mm M-61 gun for optimum aerial combat efficiency, the fighter can also carry conventional ordnance or so-called "smart" weapons without removing any of its air-to-air missiles.

WRAMA's first Eagle was disassembled by engineers and technicians to familiarize them with the F-15.

F-15s are flying all over the world today—at Okinawa, Europe and soon in Alaska.

Many new configurations of the F-15 have come about since this 1974 photo. The "G" model is currently in production.

General Hedlund welcomes the Honorable Carl Vinson to Robins. Col. Cas Conner is behind Mr. Vinson.

SAC honored the City of Warner Robins and the City of Macon by naming two B-52s. These sit side by side at Robins. The "City of Warner Robins" was the first B-52 to be stationed in Georgia.

Colonel Wys (Vice Commander) and General Hedlund welcome Madame Chiang Kai Shek in the middle 1960s.

During a campaign stop in 1964, Barry Goldwater toured WRAMA's facilities.

General Hefley delivers letter mailed through space to Representative Carl Vinson, November 22, 1960.

General Hedlund explains WRAMA exhibit at the Houston County Fair to these children standing with State Senator Ed Barker.

Armed Forces day events have always been well attended.

The XC-142A Vertical and Short Take-Off and Landing (V/STOL) aircraft managed by WRAMA tilted its wings up to get vertical lift.

Lyndon Johnson was the first President to visit Robins. With the President are Governor Carl Sanders and Representative Carl Vinson.

Upon his retirement as long term chairman of the House Armed Services Committee, Congressman Carl Vinson was succeeded by L. Mendel Rivers of South Carolina. Mr. Vinson sponsored a visit by Mr. Rivers to sustain the awareness of the committee chairman regarding the key role of Robins AFB in the national defense. Gen. Hedlund conferred with his honored guests in his office.

Dedication of marker commemorating Robins Air Force Base's Twenty-Fifth Anniversary, June 14, 1966, by Col. Bill Gibbons, Base Commander, as Gen. Hedlund, Okey Harrison, Macon Chamber President, and Jack Creamer, Warner Robins Chamber President, look on.

Governor Carl E. Sanders addressed a crowd of 75,000 at Armed Forces Day Open House celebration in 1963.

The AFLC "Horse Race" was a tool for management evaluation with the AMAs competing against each other on set criteria. This booklet contained a personal address by Gen. Hedlund which encouraged workers to continue the good work. He said ". . . victory brings a knowledge of accomplishment to every member of the team."

Senator Herman Talmadge greets a WRAMA worker.

The Thunderbirds drew large crowds for Open House celebrations.

A PHOTOGALLERY: *Leaders of the 1960s*

Commanders, Warner Robins Air Materiel Area

William T. Hefley was born in Cameron, Texas, on April 24, 1906. He graduated from the U. S. Military Academy at West Point in 1924.

Hefley's first assignment was to the District Engineer Office in New Orleans, Louisiana. From 1929 to 1931 he served with a mounted unit of the 1st Cavalry Division, at Fort McIntosh, Texas.

He completed pilot training in June 1932 and transferred to the Air Corps. His first staff assignment was in the Office of the Chief of the Air Corps in Washington, D. C. This was followed by tours of duty with the Air Service Command in Washington, as Chief of Staff of 3rd Air Service Area Command in Atlanta, Georgia, and as Deputy Chief of the Maintenance Division at the Air Service Command headquarters at Patterson Field.

In March 1944 General Hefley was assigned as Deputy Commander of VIII Air Force Service Command in England. In July 1948 he was assigned to Oklahoma City Air Materiel Area as Director of Maintenance and later became Deputy Commander.

In 1950, Hefley became Chief of Staff for Air Task Group 3.4, based at Eglin AFB, Florida. In 1951 he was assigned to Hq Air Materiel Command at Wright-Patterson AFB as Assistant for Plans, Programs, and Requirements.

From July 1953 to July 1956, he was at Sacramento Air Materiel Area, first as Deputy Commander and then Commander. In 1956 he was named Commander, Air Materiel Force, European Area, a position from which he became WRAMA Commander.

Major General William T. Hefley
November 1960—July 1963

General Hefley is married to the former Corra Lynn Dent and has one son, William T. Jr. General Hefley retired from his position as WRAMA Commander and now lives in Shalimar, Florida.

Major General Earl C. Hedlund
August 1963—July 1966

Major General Francis C. Gideon
August 1966—July 1968

Earl Clifford Hedlund was born in Valparaiso, Nebraska, on July 16, 1916. He received his Bachelor of Science degree from the University of Nebraska in 1938, and his Master of Science degree at the University of Illinois in 1939. He received his Ph.D in 1948 from the University of Illinois. He was the only commander to hold a Ph.D.

He was commissioned in June 1938 through the Reserve Officer Training Corps (ROTC) program at the University of Nebraska, and received his pilot training at Randolph and Foster Fields, Texas, graduating in 1942.

From August 1942 to 1947, he served as a fighter pilot, squadron commander, group commander, and deputy wing commmander in the Pacific and European theaters. He was credited with the destruction of fifteen enemy aircraft. In April 1945 his P-38 was shot down by ground fire. Although suffering from second degree burns, he was able to parachute from the burning aircraft only to be captured by the Germans. He later escaped and made his way back to the American lines.

From 1948 to 1952, General Hedlund was assigned to HQ USAF in the Directorate of Transportation. In 1956, he became Deputy Director of Transportation, Headquarters USAF, and in August 1959 was appointed Director of Transportation. On July 20, 1961, General Hedlund became Deputy Commander, Odgen Air Materiel Area, Ogden, Utah.

General Hedlund is married to the former Eleanor Neff of Beaman, Iowa and they have six children.

After his tour at Robins, General Hedlund became Deputy Director of Defense Supply Agency, Alexandria, Virginia. He is retired and now lives in Annandale, Virginia.

Francis Clare Gideon was born February 18, 1917 in Payne, Ohio. In 1936 he entered the U. S. Military Academy following two years study at Miami University, Oxford, Ohio. He was commissioned on June 11, 1940.

Following graduation from flying training school at Kelly Field, Texas, in March 1941, he was stationed at Langley Field, Virginia, with the 2nd Bomb Group.

In February 1942 he began a four year tour in the Pacific theater of operations. He was assigned to the Fifth Air Force's 22d Bomb Group and with Headquarters Far East Air Forces. From December 1945 to July 1951 he served various tours with the War Department, with HQ USAF and on the Joint Staff in Washington, D. C. In June 1952, he was assigned as Deputy Director of Operations, HQ MATS. For four years beginning in July 1954, he was at Dover AFB, Delaware as Commander, 1607th Air Transport Wing.

From August 1958 to February 1962 he was assigned to HQ AFLC at Wright-Patterson AFB, Ohio. There he served as the Civil Engineer, Director of Transportation, and, finally, Director of Data Systems.

He became Executive Director, Logistics Services, with the Defense Supply Agency at Alexandria, Virginia, in February 1962. In July 1964 he was assigned as Deputy Director of the organization.

General Gideon became Commander Thirteenth Air Force, Clark AB, Phillippines following his tour at Robins. He is now retired and lives in Larkspur, Colorado.

Major General A. J. Beck
July 1968—August 1972

A native of Dallas, Texas, General Beck received his Bachelor of Laws degree from Southern Methodist University, Dallas School of Law, in 1939. The following year he ended private law practice and entered the Army Air Corps.

After service in the Southwest Pacific during World War II, General Beck became one of the early pioneers of the Strategic Air Command. He commanded and brought to combat-readiness one of the early B-47 jet bomber units and served in command and staff positions at Wing, Air Division, and Air Force level and as Judge Advocate General, Inspector General and Director of Materiel for the Strategic Air Command.

General Beck came to WRAMA in July 1968 after serving two years as Senior Air Force member, Weapons Systems Evaluation Group, in the office of the Director of Defense Research and Engineering Office of the Secretary of Defense. He retired from the United States Air Force at the end of his tenure as Commander of the Warner Robins Air Materiel Area.

General Beck is married to the former Anne Michlin. They have four children. Following his retirement, General Beck served as Vice President for Development for Zales Jewelers; he resides in Dallas.

Commanders, 3AFRegion/CONAC/AFRES At Robins AFB

Lt. Gen. William E. Hall
April 1961—September 1961

General Hall was born October 22, 1907 in McAlester, Oklahoma. He graduated from West Point in 1929. He served at Albrook Field in the Panama Canal Zone and, in 1937, as Operations Officer at Bolling AFB, Washington, D. C. During World War II he served in the Mediterranean theater as Chief of the American Mission to Bulgaria and Deputy Commander of the Fifteenth Air Force in Italy. He also served in Turkey, and was in Berlin during the Berlin Airlift. He assumed command of the Fourth Air Force in 1951 and in 1952 was named vice commander of CONAC at Mitchel AFB, Long Island. He went to HQ AFRES in 1953 and in 1957 became CONAC commander.

General Hall moved the headquarters from Mitchel AFB, New York to Robins in April 1961 and prepared the Air Force Reserve's C-124 units for imminent mobilization. General Hall is retired and lives in Palm Bay, Florida.

Lt. Gen. Edward J. Timberlake
July 1962—June 1965

General Timberlake was born in Ft. Hunt, Virginia on November 10, 1909. He graduated from the U. S. Military Academy in 1931. He was assigned with the Air Corps at Kelly Field, Texas, MacDill Field, Florida, Barksdale Field, Louisiana, and the European Theater. In 1945 he went to Bolling Field, D. C. for CONAC and from there to Far East Air Forces Headquarters at Tokyo, Japan. In 1949 he was named Vice Commander of the Fifth Air Force at Itasuke, Japan and moved with it to Korea in 1950, later becoming commander.

In 1951 he was named commander of the Ninth Air Force and from there to commander of the Fourth Tactical Air Force (NATO) in 1957. In 1958 he became vice commander of USAFE in Wiesbaden, Germany and in 1961 was named Deputy Chief of Staff for Personnel, Hq USAF. General Timberlake is retired and lives in Hilton Head Island, South Carolina.

Lt. Gen. Cecil H. Childre
August 1965—May 1966

General Childre was born September 2, 1911, at Brownwood, Texas. He attended Southwestern University at Georgetown, Texas and Huntingdon College, Montgomery, Alabama. He served at Kelly Field, Hondo Army Air Field, and Brooks Field in Texas. In 1947 General Childre was assigned to the Air Command and Staff School at Maxwell AFB, Alabama. In 1950 he went to Japan with the Fifth Air Force and from there to Korea as vice commander of the 315th Air Division under FEAF. In 1953 the General was assigned to Donaldson AFB, South Carolina. In 1954 he went to Langley AFB, Virginia for TAC. In 1960 he became Assistant Deputy Chief of Staff for Plans and Programs at HQ USAF. His last assignment before coming to Robins was at Ankara, Turkey. General Childre died on 28 May 1966.

Lt. Gen. Henry Viccellio
August 1966—July 1968

The Air Force Reserve units mobilized on two occasions in 1968, in January, following the *Pueblo* attack, and in May to reinforce the Strategic Reserve. It was General Viccellio's duty to discontinue HQ CONAC on 1 August 1968. General Viccellio is deceased.

Maj. Gen. Rollin B. Moore, Jr.
August 1968—January 1972

General Moore was born in Chico, California, on June 3, 1917. He attended the University of California. He was commissioned in August 1940 and served on active duty from 1940 to 1946 at Stockton Army Air Field, California; Douglas Army Air Field, Arizona; and Pecos Army Air Field, Texas. He served with the Air National Guard at Alameda, Hayward, Fresno, and Hamilton AFB, California. In 1960 he was picked to command all Air Force Airlift Forces in the biggest Joint Army-Air Force exercise of that time.

General Moore is married to the former Elaine Eckenroth and they have two children, Rollin Joseph and Patricia Lynn. He is retired and living in Orinda, California.

Biographical Notes on the Commanders, 3AF Region/CONAC/

Col. Harry S. Bishop, November 1960—April 1961 (not pictured): Colonel Bishop is a native of Plainview, TX. He graduated from the U. S. Military Academy in 1933 and was commissioned in the U. S. Army. He served at Kelly Field and Lubbock, TX; and in Eritrea, British East Africa, India, Turkey, and Europe. He previously served at Robins as Vice Commander of the Fourteenth Air Force. Colonel Bishop moved the region headquarters to Dobbins AFB, Georgia, during March 1961 to make room for Headquarters CONAC which was to move down from Mitchel AFB, NY.

Lt. Gen. Gordon A. Blake, October 1961—June 1962 (not pictured): General Blake was born in Charles City, Iowa, on July 22, 1910. He graduated from the U. S. Military Academy in 1931. He served with the Air Corp at Barksdale Field, LA, Fort Monmouth, NJ, and Chanute Field, IL. In 1939 he went to Hawaii and from there to the Philippines. He was serving at Hickam Field on December 7, 1941 and was awarded the Silver Star for gallantry in action that day. He served as vice commander at Wright Field from 1952 to 1953. In 1953 he went to HQ USAF and from there to command the United States Air Force Security Service in San Antonio, TX. In 1959 he was assigned as Vice Commander in Chief of Staff for PACAF. General Blake is retired and lives in Universal City, TX.

Deputy/Vice Commanders, Warner Robins AMA

Col. Castex P. Conner
July 1960—July 1964

Col. Robert A. Wys
August 1964—December 1967

Col. Edwin J. Hamilton
December 1967—June 1968

Col. James A. Bailey
June 1968—May 1969

Col. Ralph A. Johnson
May 1969—August 1969

Brig. Gen. Ralph T. Holland
August 1969—November 1971

Biographical Notes on the WRAMA Deputy/Vice Commanders

Colonel Conner: Castex Paul Conner was born in Jennings, LA on 31 October 1916. He graduated in June 1938 from the U. S. Military Academy, West Point, NY. He served with ATC at several bases; at the Oklahoma Air Depot; in the Marianas Islands with the wing responsible for dropping the first atomic bomb; in the Philippines and in Japan; with the faculty of the Industrial College of the Armed Forces; with HQ PACAF; and at the Daytonl Air Force Depot. After his retirement from the Air Force at WRAMA, he and his wife moved to Macon, where they continue to live following his retirement from the position of Vice President of the First National Bank in that city.

Colonel Wys: Colonel Robert A. Wys is a native of Peoria, IL. A graduate of Knox College, he received his commission in 1939. He served in the China-Burma-India theater; with HQ ATC at Barksdale AFB, LA, and Mitchel AFB, NY; with CONAC at Mitchel AFB, NY, and Colorado Springs, CO; with USAF and OSAF at the Pentagon; at Toul Rosieres Air Base; with HQ AFLC at Wright-Patterson AFB, OH and at Robins. After retiring from the Air Force in 1967, he served on the staff of Wesleyan College in Macon on developmental matters, a position from which he has retired. He lives in Macon.

Colonel Hamilton: "Ed" Hamilton is a native of Delta, Iowa. He is a 1937 graduate of Iowa Wesleyan college. He began his Air Force career, which was chiefly in the supply and maintenance fields, in December 1940 as a flying cadet. He came to Robins in August 1963 as Director of Materiel Management from Alexandria, VA, where he was Chief, Inventory Division, Detachment 10, Defense Supply Agency. He served as Deputy Commander, 2843d Depot Wing, Olmsted AFB, PA, and as Director of Maintenance Engineering and as Director of Supply and Services, 2844th Depot Wing at Dayton AF Depot, OH. His overseas service includes three years in Japan and three in England. Prior to his assignment to DSA, he was with the office of the DCS/Materiel, Headquarters USAF. His wife is the former Miss Betty Covert of Rumsey, CA. Their two daughters are Cheryl Lynn and Gayle Anne. He is retired and living in Florida.

WRAMA Executive Assistants/Chiefs of Staff/Directors/Deputies and Tenant Commanders:

Mr. George O. Cornish
Executive Assistant
September 1964—January 1969

Mr. Shaffer T. Day, Jr.
Executive Assistant
May 1969—September 1969

Col. Ralph A. Johnson
Chief of Staff
October 1969—August 1970

Col. James W. Thompson
Director of Supply and Transportation
April 1961—July 1965

Col. Elmer H. Epperson
Director of Supply and Transportation
July 1965—November 1968

Col. R. H. Worrell, Jr.
Director of Distribution
October 1968—April 1970

Col. Morgan W. Pirkle
Director of Maintenance
September 1960—September 1962

Col. John W. Hope
Director of Maintenance
September 1962—July 1964

Col. Maurice E. Parker
Director of Maintenance
July 1964—November 1968

Col. William O. Bailey
Director of Maintenance
November 1968—October 1969

Col. W. R. Kelso
Director of Maintenance
October 1969—April 1972

Mr. Shaffer T. Day, Jr.
Deputy Director of Maintenance
May 1966—April 1970

Col. Richard W. Dyer
Director, Materiel Management
August 1961—July 1963

Col. Edwin J. Hamilton
Director, Materiel Management
August 1963—December 1967

Col. James A. Bailey
Director, Materiel Management
December 1967—May 1968

Col. Howard G. Beck
Director, Materiel Management
June 1968—August 1970

Col. Robert E. Lee
Director of Procurement and Production
July 1961—June 1963

Col. Harry Zipper
Director of Procurement and Production
July 1963—July 1966

Col. Ralph A. Johnson
Director of Procurement and Production
August 1966—May 1969

Col. Charles W. Delanoy
Director of Procurement and Production
August 1969—June 1970

Col. Samuel F. Langley
Chief, Plans and Management Office
July 1961—March 1963

Col. William J. Atkinson
Chief, Plans and Management Office
July 1963—July 1967

Col. Howard G. Beck
Chief, Plans and Management Office
August 1967—May 1968

Col. Joseph E. Cahill
Director, Plans and Programs
June 1968—June 1970

Mr. Earvin L. Joyner
Deputy Director, Plans and Programs
December 1969—September 1975

Col. Clifford V. Warden
Base Commander
December 1961—July 1963

Col. Robert A. Shafer
Base Commander
September 1963—May 1964

Col. William E. Gibbons
Base Commander
July 1964—June 1967

Col. Andrew A. Chaffin
Base Commander
July 1967—October 1969

Col. Frank W. Metheney
Base Commander
October 1969—June 1973

Col. John C. Healey
Director of Personnel
September 1965—January 1968

Col. Gordon K. Stallings
Director of Personnel
January 1968—December 1969

Col. Lester C. Mourer
Comptroller
July 1962—June 1964

Col. Alexander J. Almand
Comptroller
June 1964—February 1969

Col. Grady D. Gafford
Comptroller
April 1969—July 1973

Mr. Stephen J. Chasse
Deputy Comptroller
March 1962—1979

Col. Billy J. Millis
5th Mobile Communications
Group Commander
August 1965—June 1967

Col. Joseph L. Solomon
5th Mobile Communications Group Commander
June 1967—July 1969 and
August 1970—July 1971

Col. Lloyd N. Bergman
5th Mobile Communications
Group Commander
July 1969—August 1970

Biographical Notes on the WRAMA Executive Assistants/ Chiefs of Staff/Deputies and Tenant Commanders:

Mr. George O. Cornish: Mr. Cornish came to Robins from the Ogden AMA. He is a graduate of the Air War College and has received numerous awards including the Ogden AMA Commander's Award in 1959. Mr. Cornish left Robins to accept a postion with the Defense Supply Agency in Cameron Station, VA.

Col. James W. Thompson: Colonel Thompson was personally involved with installing the first conveyor lines in and enlarging the Air Freight Terminal to accommodate increased cargo shipments by air. He retired from this position and lives in Spartanburg, SC.

Col. Elmer H. Epperson: Colonel Epperson was personally involved in the preliminary design for a storage facility for small items. Retiring from this position at WRAMA, he accepted a position in the aerospace industry in Fort Worth, TX.

Col. R. H. Worrell, Jr.: Colonel Worrell led in the construction and completion of the small items facility, which featured a magnetic in-floor tow-cart system and a monorail conveyor system with photo-electric coding for directing materiel to proper locations. He retired from this position at WRAMA and moved to Henderson, NC.

Col. Morgan W. Pirkle: As the overhauling of R-4360 engines phased out, a training program was initiated in electronics to meet advances in technology associated with airborne navigation/communication/radar/fire control systems. Upon his retirement from active duty, Colonel Pirkle served in civil service as Deputy Chief of the Shops and Resources Management Divisions in the Directorate and was active in the civic affairs of the City of Warner Robins. He now lives in Amelia Island, FL.

Col. John W. Hope: Under his leadership, Maintenance was assigned Specialized Repair Activity (SRA) for B-57, C-130, and C-133 aircraft and components as well as SRA for airframe components of B-47, C-97, DC-97, and C-124 aircraft. Robins was noted for airborne armament repair during this period. Colonel Hope is retired and lives in Fort Lauderdale, FL.

Col. Maurice E. Parker: The first C-141 Starlifter came to Robins in this time period; the aircraft paint shop was enlarged to accommodate one C-141A or two C-130 aircraft. Under his leadership, the Depot Maintenance Industrial Fund was effected on July 1, 1968 to function as a customer sales operation with customers such as Air Force, Army, Navy, and others. Colonel Parker is retired and lives in Hilton Head, SC.

Col. William O. Bailey: During this period the C-130 and AC-119 Gunships were modified. These aircraft were used for support to Southeast Asia.

Col. W. R. Kelso: A major project during this period was the production of two prototypes with follow-on production of six AC-130E PAVE SPECTRE gunships, incorporating advanced technology and sophisticated equipment. Under Colonel Kelso's leadership, the directorate received top honors as the best work center in AFLC for FY 1970. Colonel Kelso is employed by Hughes Aircraft Corporation in Los Angeles.

Mr. Shaffer T. Day, Jr.: Mr. Day served as Executive Assistant to the WRAMA Commander prior to being assigned as civilian assistant to the Air Force deputy chief of staff of systems and logistics in Washington D.C. Since retiring, Mr. Day has been employed in private industry and has represented Lockheed Aircraft Corporation in Greece. He is now retired in Ontario, CA.

Col. Richard W. Dyer: He retired following his service at Robins AFB and resided in Florida. Colonel Dyer is deceased.

Col. James A. Bailey: Following Colonel Bailey's service at Robins, he rose to the rank of Major General as Comptroller, Air Force Logistics Command. Following his retirement from the Air Force, he became Vice President of the Seaboard Coast Line Railway, in charge of data processing. He has since retired from that position and remains in Jacksonville, FL.

Col. Howard G. Beck: Colonel Beck retired at Robins from the position of Chief of Staff of the Warner Robins Air Materiel Area. Following a period of employment in men's fashion retailing in Jacksonville, FL, he is now employed with a major jewelry retailer headquartered in Dallas, TX.

Col. Robert E. Lee: Colonel Lee left Robins to become Deputy to the Director of Procurement Policy, Deputy Chief of Staff, Systems and Logistics, Headquarters USAF, Washington D.C. He retired at the rank of Brigadier General and lives in Pensacola, FL.

Col. Harry Zipper: During this period, the directorate was responsible for purchases of equipment, services, and supplies amounting to some $500 million annually. Colonel Zipper left Robins to become Commander of the Air Procurement Region, European, at Chateauroux Air Station, France.

Col. Ralph A. Johnson: During Colonel Johnson's tenure, the firm-fixed-price contract system was adopted and the government was encouraged to prepare more precise Statements of Work. By 1968, fixed-price contracts represented seventy-eight percent of defense procurement (a thirty-five percent increase over 1960). Many of WRAMA's systems, modifications and spares were purchased by these new methods—including the F-15. Colonel Johnson served as Executive Assistant to the WRAMA Commander until his retirement in 1970. He is currently serving as Executive Vice President of the Warner Robins/North Houston County Chamber of Commerce, maintaining close ties with Robins Air Force Base.

Col. Charles W. Delanoy: Colonel Delanoy is retired.

Col. Samuel F. Langley: Colonel Langley left Robins to become head of the Defense Industrial Plant Equipment Center, Memphis, TN. He retired from active service in October 1965. For many years he served ably as an executive with a civic organization seeking industry in Memphis, TN; he then served as head of the Memphis/Shelby County Hospital Authority. He is now retired and lives in Memphis.

Col. William J. Atkinson: Colonel Atkinson retired in February 1968 and lived in Warner Robins; he is deceased.

Col. Joseph E. Cahill: During his tour, the office was changed to the Directorate of Plans and Programs. Colonel Cahill left Robins for an assignment at HQ USAFE in Wiesbaden, Germany. He is now employed by the State of Florida and lives in Tallahassee.

Mr. Earvin L. Joyner: Mr. Joyner led in the planning which made Warner Robins the avionics center of the Air Force; which diversified the Center from cargo aircraft specialists into a broad spectrum Center with the assignment of the F-15; and which established the major facility/equipment modernization program which is still underway. He left XR to become the Deputy Director of Maintenance, a position which he still holds.

Col. Clifford V. Warden: During Colonel Warden's tour CONAC Headquarters was transferred from Mitchel AFB, NY to Robins.

Col. William E. Gibbons: Under Colonel Gibbons, the Robins Park Area was completed and the Base Recreation program was greatly improved. Following his retirement from Robins, he invested in real estate in the Macon area and now works and lives at Lake Wildwood.

Col. Andrew A. Chaffin: Colonel Chaffin was instrumental in establishing Gate #1 and Green Street as a major traffic artery. Colonel Chaffin has retired and now lives in Gadsden, AL where he is an executive with the Alabama State Highway Department.

Col. Frank W. Metheney: During his tenure a great improvement was made in Base Support Facilities as The Community Center began to take shape with the completion of the Base Exchange, Theater, Post Office, and Bank.

Colonel John W. Hope, Chief, Personnel and Administration, August 1961—December 1961 (pictured with Maintenance Directors).

Colonel Earl D. Duncan, Director, Personnel and Administration, December 1961—June 1964 (not pictured): Colonel Duncan was reassigned to Griffiss AFB, NY.

Col. Phillip P. Grove, Director, Personnel and Administration, June 1964—September 1965 (not pictured): Colonel Grove retired in 1965.

Col. John C. Healey: He was reassigned as Chief of the Information Office.

Col. Gordon K. Stallings: He became the WRAMA Inspector General in 1969.

Col. A. N. Richard, Jr., Director of Personnel, December 1969—16 July 1970 (not pictured): Colonel Richard completed a detailed study of actions necessary to establish an on-base four-year baccalaureate degree-granting program oriented toward a bachelors degree in logistics management. He lives in Atlanta.

Col. Lester C. Mourer: Colonel Mourer was reassigned to HQ PACAF.

Col. Alexander J. Almand: Colonel Almand retired in April 1969 to pursue a career as a civilian banker in Macon, GA.

Col. Grady D. Gafford: Colonel Gafford retired from his position as WRAMA Comptroller in July 1973.

Mr. Stephen J. Chasse: During this time the Data Automation activity continued to grow as customer needs increased. Two IBM 7080s (large scale data processing systems) were installed along with five 1401 EDPS medium scale computers to serve as input/output data processing devices. Five RCA 301 computers (medium scale computers) were installed and in 1965 equipment was dedicated for exclusive processing of selected data systems. In 1970 the first multiprocessor computer B3500 was installed and later a larger scale multiprocessor B4700. Other large scale and dedicated computers were received and made this one of the largest computer centers in the southeast. Mr. Chasse is retired and lives in Warner Robins.

Col. Wesley L. Pendergraft, 19th Bomb Wing Commander, July 1968—April 1969 (not pictured): Colonel Pendergraft was the first Commander of the 19th Bomb Wing at Robins.

Col. G. R. Abendhoff, 19th Bomb Wing Commander, April 1969—February 1970 (not pictured): Colonel Abendhoff retired in August 1974.

Col. Billy J. Millis: A significant achievement during the first few months was the rehabilitation of several World War II structures for use as office and living space for the group. These structures were made habitable through a self-help program initiated by Colonel Millis. During this period, two "Talking Bird" airborne communications packages and advisory support responsibility for ten Air National Guard units were acquired by the group. Also, Detachments one and two were activated at Key West and McCoy AFB, FL. Col. Millis is retired from military service.

Col. Joseph L. Solomon: During Colonel Solomon's tenure the *Bare Base* deployment concept and the Air Force Component Command Post were established. Also, an open storage area was obtained through self-help for vehicles and other equipment assigned to the 5th to support the *Base Base* concept. A Chevrolet Stepvan was also acquired to house "Talking Bird" communications-electronics equipment. Col. Solomon is retired and lives in Warner Robins.

Col. Lloyd H. Bergman: The group deployed eight personnel and four vehicles to aid in the relief effort of residents on the Gulf Coast after Hurricane Camille. A C-130 "Talking Bird" communications package was sent to Gulfport, MS Municipal Airport to provide communications for coordinating airlift of critical supplies into disaster areas. The operation proved that tactical communications and air traffic control equipment could be deployed and set up on a *Bare Base* in a matter of hours. Colonel Bergman is now retired.

CHAPTER FOUR: *The 1970s: In New Directions*

The Southeast Asia Conflict Continues

WRAMA People, Planes and Parts

During the early 1970s, as the Southeast Asia situation intensified, WRAMA and its people continued to invest heavily in the effort. This investment was not only in supplies, such as aircraft, spare parts, and massive quantities of sophisticated electronic warfare equipment, but in WRAMA personnel who voluntarily went to Vietnam to establish and operate storage and repair facilities in such places as Da Nang and Bien Hoa. Leaving families behind, WRAMA people relocated into almost unbelievably frenetic surroundings, not totally unlike moving into an ant colony. Facilities were sparse and somewhat reminiscent of World War II in the South Pacific: Quonset huts, tents, palm trees, and the ever-present sound of aircraft.

WRAMA people saw themselves as more deeply involved in the war than any other AMA. At one point, WRAMA had twenty-eight primary systems operating in Southeast Asia and was providing support to more than half of the U. S. aircraft in Vietnam. More than eighty percent of the Vietnamese aircraft were managed and logistically supported by WRAMA.

At home, efforts continued to devise exotic, yet potent, weapons to overcome some of the problems of air-to-ground gunnery being faced in fighting limited, guerrilla wars. The AC-47 gunship, "Puff, the Magic Dragon," which was the forerunner of the AC-119 and AC-130 gunships, was mighty, indeed, but a mere token of the firepower to come. The AC-130s, which were both managed and modified at WRAMA, were seen as instant destruction with wings.

Special Weaponry

The Air Force gunship program, which began with the AC-47 modifications in the middle 1960s, progressed through the more potent AC-119G/K gunships of the late 1960s. Gunship capability, developed for the most part in response to operational contingencies in Southeast Asia, increased dramatically during the late 1960s and early 1970s. By 1973, they had become heavily armed, side-firing weapons platforms equipped with complex arrays of navigation, communication, and sensory equipment that had the ability to locate and destroy enemy personnel, vehicles, and supplies in nighttime surprise attacks.

AC-47, "Puff the Magic Dragon" Gunship, forerunner of the mighty AC-130 Pave Spectre Gunships.

In 1970, ASD initiated development of the prototype of the AC-130E *Pave Spectre I* gunship. The work was done in-house at WRAMA, with WRAMA filling the contractor's role and ASD acting as Program Manager. The AC-130Es incorporated several advances in gunship technology, including a digital computer, a new fire control system, a laser range designator, electronic countermeasures (ECM) gear, and improved infrared and navigational systems.

The SEA operations continued to spur gunship development. In July 1972, WRAMA was given the task of producing three additional AC-130 gunships, which were to include several major improvements and which were to be identified as *Pave Spectre II*. Although the basic airframe remained the same, the addition of more powerful engines caused the aircraft to be redesignated as AC-130Hs. Firepower was also increased by the addition of the 105 mm trainable gun, of higher caliber than previous armament, with an electro/hydraulic system that enabled the gun to adjust to the precise position of the target. This reduced the requirement for "aiming" the aircraft. The 40 mm gun already installed on the AC-130H was also equipped with the electro/hydraulic system.

"Puff" Armament—three miniguns, each capable of firing 6,000 rounds per minute.

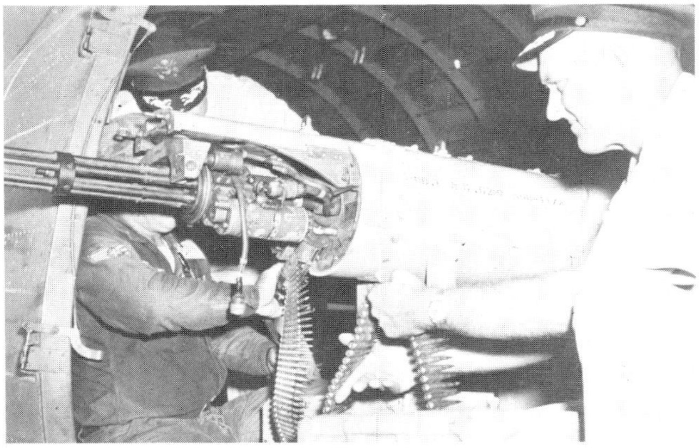

Feeding "Dragon Teeth" to the miniguns on the "Dragon."

Business end of the AC-119 Gunship—four miniguns; another forerunner of the AC-130 Gunships.

AC-130, Pave Spectre I Gunship with "Black Crow" sensor (note small radome just above the nose wheels). This enabled it to hunt down enemy supply trucks in the dark by homing in on the radio frequency emissions of the trucks' ignition systems.

One 40 mm gun was eliminated and the APQ-150 beacon tracking radar was located in its place. The chaff and illuminator dispenser was moved to a new wing pylon, allowing the rear cargo door/ramp to be closed, thus significantly reducing drag and undesired airflow inside the fuselage.

The Low Light Level Television (LLLTV) was further improved by utilizing a laser to provide better target resolution, and observation bubbles were added to the aft and right scanners' positions to improve their field of vision.

The fire control system of both *Pave Spectre* models utilized a digital computer to integrate functions of the multiple systems which eliminated the necessity for nominal aircraft orbit, consisting of specified bank angle, altitude, and air speed. The result? A real war wagon that operated with the smooth efficiency of a modern automatic camera—just point and shoot.

Continuing to Build

Within the WRAMA management/organizational structure, selected individuals carried much of the responsibility for the logistical support to SEA. The System Support Manager (SSM) was charged with maintaining support for the assigned weapon system throughout the world, and the Item Manager (IM) was responsible for providing certain types of supplies and equipment. Perhaps less known was the Project Manager, a middle manager who was the key to the success of many operations and was the coordinator of all actions within a command for the support of a given project. Some individuals were assigned this responsibility for several projects. In many respects, these were the unsung heroes of the Vietnam logistical support operations.

Another WRAMA program in 1971, designed to increase the Vietnamese Air Force's independence as part of the Vietnamization effort, was the preparation of two-language technical manuals which would allow the Vietnamese to do much of their own equipment repair. These simplified two-language publications, called "Job Performance Aids" (JPAs), used special technical language, illustrations, and Viet-English dual instructions, and some proponents felt that the new techniques employed would be a breakthrough in spreading technological know-how to underdeveloped nations. The project involved the cooperation of field forces, ASD at Wright-Patterson AFB, Ohio, WRAMA personnel, contractors in Alabama and California, and the U. S. Army.

JPAs were planned for South Vietnam to cover maintenance on the CH-47 Chinook, the C-123K Provider, the C-7A Caribou, the R-8 aircraft refueler, the 530B and 0-11A firetrucks, and the UH-1H Iroquois Helicopter.

To accommodate the fact that many of the technical terms did not exist in the Vietnamese language, and to work around the multiple dialects spoken in Vietnam, a standardized technical language dictionary was developed containing some three thousand words. This was of great assistance to both human and computerized translators.

Although the technical manuals involved both Air Force and Army technologies and techniques, they were consolidated into small, handy books of simple instructions, in absolutely logical order, telling the worker what to do, where to do it, and most importantly, how to do it in terms and pictures he could understand.

Prior to the 1970s, transport aircraft, primarily the C-130 in the 1950s and the C-141 in the 1960s, had been symbolic of the WRAMA prime weapon system responsibilities. In fact, virtually every type transport and helicopter in the Air Force inventory came under WRAMA management sometime during this period. Many of these were the most useful and effective aircraft in the Air Force's combat operation in the lengthy Vietnam struggle, and most of them remained a WRAMA/WR-ALC responsibility throughout the 1970s.

The decade of the 1970s, however, was to introduce to the logisticians at Warner Robins the first fighter aircraft ever assigned to the Middle Georgia center as one of its prime management responsibilities—the F-15 Eagle, America's newest air superiority fighter. Although announcement of the management assignment was made in 1969, the first Eagle did not officially reach WR-ALC until May 1975. It would, however, become one of the center's most prestigious management and maintenance assignments.

Vietnam Aftermath and Mission Changes

At the end of the Southeast Asia Conflict, the mass migration of Vietnamese and other Asians from Vietnam, Cambodia, and Thailand affected many USAF installations from Clark AFB in the Philippines all the way to Robins Air Force Base.

Many of the refugees were brought to Eglin Air Force Base, Florida and, of course, needed food, clothing, and other supplies. Robins Air Force Base contributed eleven truckloads of supplies that were moved on trucks brought to Robins from Dalton, LaGrange, and Albany, as well as many locations in the Middle Georgia area. The trucks were loaded with 142,229 pounds of cargo and were moved to Eglin Air Force Base on April 30, 1975. This movement of supplies proved to be of great assistance in housing and feeding the thousands of refugees temporarily housed at Eglin.

Along with the phase-down of support to Vietnam came numerous changes in the WRAMA mission. The principal effort, once devoted to providing support for our troops in Southeast Asia, now became a push towards survivable workloads, those systems and end items that prepare the Air Force for a more rapid response to future contingencies, should they arise, and yet remain active during peacetime.

The F-15 Eagle was one of these workloads, and it was quickly recognized that it would be one of our most effective weapons for years to come.

Under the Program Management Responsibility Transfer (PMRT) Plan, full management responsibility for the Eagle will be transferred to the Air Force Logistics Command (AFLC) in the fall of 1982, with the airframe, avionics, radar armaments, and the TEWS assigned to WR-ALC. This will assure a major workload for this ALC's maintenance shops, storage, and system/item managers well into the future.

Inflation Settles In

As the Air Force's, and the nation's role in Southeast Asia began to subside, more money became available for use at home and, little by little, salaries and market prices began to rise. The higher salaries were passed on to the consumers through higher prices for almost everything—food, clothing, housing, interest rates—followed by a cry for higher salaries to compensate for the increased cost of necessities and luxuries. It began to be a vicious circle, and the inflation race was on.

Within the government, the push was to adjust for higher salaries by using fewer people in an attempt to live within congressional appropriations. While the defense budget remained relatively stable from 1968 through 1976, the impact of inflation reduced the real buying power. At the same time, sophistication and complexity of modern weapons continued to grow, and so did the costs. Additionally, other world situations plus the residual "scare" effect from the Vietnam conflict heightened the Department of Defense's desire to be better prepared "next time." Readiness efforts increased workloads, even though manpower was being reduced. This, in turn, made more work per person, causing additional desires for increased pay. This inflation has been, in recent years, the biggest problem to confront the installation—and the nation—and has not yet been dealt with satisfactorily.

Just as all of this inflation and availability of money was about to peak, or so everyone thought, along came a shortage of energy supplies. Due to the Arab oil embargo, and our own dwindling supplies, petroleum fuels weren't as readily available on the market, at any price. As a result even more dollars became available for other things, bringing about several years of two-digit inflation.

Doing More with Less— The Energy Crisis

Glad to be out from under the pall of the Southeast Asia conflict and ready to go places and do things, the American people were to face a new and radically different world of energy shortages. Diminishing oil reserves caused oil to become scarce—and of course, expensive. Little did anyone realize just how expensive it would become later on. With the reduced energy supply, the WRAMA thrust became "doing more with less" and "work smarter, not harder." The watchword became *"Productivity,"* and innovation would soon become the only way to survive—finding better work methods, improving our facilities, and cutting out nonessentials.

Crested Rooster

The HH-53 Super Jolly Green Giant is the prime launch site astronaut recovery vehicle in the space program, and is also capable of spacecraft recovery.

In early 1973, the Air Force directed the modification of six HH-53 helicopters for use in the recovery of unmanned space hardware within an eight hundred nautical mile radius of Hickam AFB, Hawaii. The HH-53s were to be supported by three C-130 tankers. Although the cost of modification was in excess of $1.5 million, the use of HH-53s and C-130s would save more than $2 million per year in operational costs over the existing method of using two converted Liberty ships and six CH-3B aircraft. The modifications to the HH-53s consisted of installation of a heavy duty crane, two additional fuel tanks, a passenger comfort cubicle, and a scuba equipment storage area. The project, code-named *Crested Rooster* was completed in June 1974—one year sooner and $3.5 million cheaper than the contractor's bid. It was the first large project ever undertaken by this depot in which it had complete design responsibility as well as responsibility for the manufacture and installation of the hardware.

Unfortunately world situations didn't realize they were supposed to cooperate. In October 1973, the Israeli war with Egypt caused a frantic, although well organized, response from WRAMA. During the Middle East conflict's one month duration, WRAMA shipped 4,777 tons of materiel to Israel. Identified by the name Project *Nickel Grass*, this materiel totalled $33.5 million in missiles, aircraft spare parts, ground equipment, and electronic countermeasures equipment. Also, twelve C-130E aircraft were shipped which cost an additional $32.4 million. During this same time period, the WRAMA Air Terminal processed and shipped an additional five thousand tons of materiel from other agencies such as the Army, the Marines, etc., which were trucked or flown into WRAMA to be aggregated and loaded onto the aircraft which would carry the supplies to Israel.

During a visit to Robins AFB in November 1973, President Nixon particularly commended WRAMA and said, "I want to compliment you on the outstanding job your people did on the airlift to the Middle East."

On November 18, 1973, President and Mrs. Richard M. Nixon came to the Macon/Warner Robins area to honor Representative Carl Vinson on his Ninetieth Birthday, and to participate in the One Hundredth Anniversay Celebration of the Walter F. George School of Law at Mercer University in Macon, Georgia.

President Nixon was greeted by 28,000 people upon his arrival at Robins Air Force Base.

At ceremonies in the Virginia shipyards, President Nixon presented a model of the nuclear powered aircraft carrier USS Carl Vinson *to Mr. Vinson. It is the first major American warship to be named after a living person. Secretary of the Navy John Warner and Secretary of Defense Melvin Laird share in the event.*

The United States' military assistance to Israel also brought an immediate response from the Arab oil-producing nations. Within one week, the Arabs doubled, then quadrupled, the price of a barrel of oil. Then on October 21, 1973, they imposed a total ban on oil exports to the United States. This embargo remained in effect until middle March 1974. Although the embargo was lifted then, the cost of oil remained at its high price, and further increases occurred in the years which followed.

The Arab Oil Embargo had a profound effect on all oil-importing nations. It dramatically revealed how dependent the individual nations are on oil imports. Within a short span of time, fuel shortages had become apparent, and many U. S. industrial facilities were forced to either reduce or curtail their production, leading to increased unemployment, higher prices, and contributing to inflation above ten percent. The increased price of the foreign oil also seriously affected the U. S. foreign balance of money payments, since there was vastly more money going out than coming in. Of further concern was the constant threat that the oil-producing nations might reimpose their oil embargo, even more seriously affecting the economy and defense capabilities.

It was obvious that the impact of the great energy crisis, which was affecting virtually all the nations of the world, was certainly making life different for the people of Robins AFB. WRAMA was faced with one of the most demanding management efforts it had ever encountered. By middle October 1973, WRAMA had developed and published the WRAMA/RAFB Energy Conservation Plan calling for a ten percent reduction in consumption of energy.

The seriousness of the situation finally began to reach everyone at Robins AFB by middle December 1973, and suggestions were received calling for consideration of a four-day work week, elimination of swing and owl shifts, changing day shift hours, reduction in lunch periods, elimination of time clocks, reduction or elimination of hot water in the restrooms, reduction of lights in nonworking areas, establishment of a bus system to areas where employees lived, preferred parking for carpools, establishment of bicycle paths, and moving the janitorial shift from night to day. Although only a few of these conservation suggestions were actually implemented, hard work and a mild winter that year allowed the base to reduce energy consumption by 17.2 percent.

Suggestion Program

Almost forty years ago, September 10, 1943, the Air Service Command, initiated the Suggestion Program. Shortly afterward, the Warner Robins Depot implemented the program.

The Command's adoption rate for its first full year was thirteen percent and cost reductions were valued at $10,006,633. During the first quarter of Fiscal Year 1982, the Warner Robins Air Logistics Center had cost reductions under the Suggestion Program of $8,648,750. The Suggestion Program is "big business" at Robins.

Another example of doing more with less evolved from the Middle East Conflict, when C-141 flights were required from the east coast of the United States to Israel to keep the Israeli armed forces resupplied. For some time, the Air Force had known that the C-141 transport's lifting capability far exceeded its fuselage volume capacity. When originally designed and built, the Air Force planners intended for the C-141 to supply European operations nonstop from United States bases on the Atlantic Coast, or to supply bases in the Asian theater after a refueling stop in Hawaii. However, flights to the Middle East demanded much more range than those to Europe, and a politically stable refueling point did not exist along the way.

The long and the short of it—the "Stretch" C-141 is in the foreground. Its added length allows it to carry thirty-three percent more cargo than the standard C-141 beyond it to the right.

Both problems were solved by the C-141 Stretch Modification which was designed to increase the airlift capability of the United States and permit timely deployment of fully equipped division-sized forces to Europe and elsewhere. This modification added 280 inches to the fuselage length and increased the transport capacity by 33 percent, equivalent to adding 90 new C-141 aircraft to the fleet at one-fourth the cost. Additionally, since the Mid-East Conflict had pointed out the dependency on overseas bases, such as Lajes in the Azores, to support long-range airlift operations, the C-141s were also modified for aerial refueling to eliminate the requirement for ground servicing enroute. The addition of the inflight fueling receptacle gave the C-141 an almost unlimited range. Since the stretch modification was a reconfiguration of the same aircraft, rather than an entirely new aircraft, it had a relatively insignificant impact on operating and support costs, relying on resources already in existence. The program was completed in June 1982, ahead of schedule and under cost.

Although Robins AFB had met and exceeded the Air Force and AFLC energy conservation goals during the 1970s, it was realized that innovative advances would have to be made during the 1980s.

In FY 1980, an Energy Monitoring and Control System (EMCS) was installed, at a cost of approximately $2 million, which allows central control for manual and programmed turn-on and turn-off of air conditioning, heat, and other equipment in all major RAFB facilities. The EMCS provides monitoring of steam pressure, chilled water temperature, airflow, metering capability, and preventive maintenance information, and is playing a major role in meeting energy goals at Robins AFB during the 1980s.

Robins AFB also succeeded in obtaining $1.5 million for FY 1980 energy saving projects for a variety of buildings as a result of the Building Energy Audit Program. These projects included low technology conventional energy saving measures such as adding or improving insulation and storm windows, as well as technical advances in energy efficiency through improved heating and cooling systems and the construction of a solar water heater.

Separatory funnels in the physical sciences laboratory.

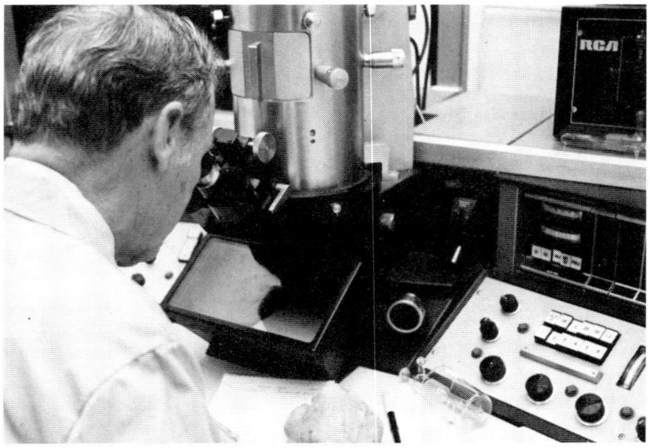

Metallurgist using a transmission electron microscope for metallurgical failure analysis.

Physical Sciences Laboratory

From a test tube operation with limited capabilities in an inspection division to a multi-faceted, versatile industrial laboratory today, is the success story of the Warner Robins Air Logistics Center's Physical Sciences Laboratory.

Known as the Chemical Lab in the late 1940s and located in the Inspection Division of the Directorate of Maintenance, it had a six-person staff. Its name was changed to Chemical and Materials Laboratory in 1954 and to its present name of Physical Sciences Laboratory in 1966. Its personnel grew to a current 42-person staff.

At the outset, the Chemical Lab's main assignment was to control electro-plating and cleaning solutions through analyses and to recommend what was necessary for their upkeep. In 1954 the lab gained non-destructive inspection responsibilities as well as metallurgical capabilities, mainly associated with support to the heat treatment and foundry shops.

In 1963 the lab began a Spectrographic Oil Analysis Program (SOAP) on all aircraft engines assigned to the Warner Robins Air Materiel Area. SOAP predicted which part of the engine may have gone bad before the engine failed, thus permitting more precise scheduling of inspection intervals and more specific repair and overhaul instructions.

Five years later the Physical Sciences Lab became more diversified. It broke away from the test tube type analysis and began instrumental analysis, allowing a much faster response. This was important to the WRAMA operation since the lab's value hinged on its ability to respond rapidly to problems.

Over the years the lab added the electron microscope, X-ray spectrograph, X-ray defraction, electron probe micro-analysis, and infrared spectrometer.

Chemists and technicians analyze, test, and evaluate paints, organic coatings, adhesives, and sealants. They analyze and control cleaning, stripping, and plating solutions used in the shops and develop new chemical processes to improve production methods. They test fuels, lubricants and aviators' breathing oxygen.

The Methods Laboratory functions as a part of the Physical Sciences Lab team. Mechanical engineering and electronics technicians look for new repair procedures for parts considered throwaway, which cost relatively little years ago but today may be not only very expensive but no longer available.

Located in Building 165, the Physical Sciences Lab supports all WR-ALC directorates and tenants as well as customers within the center's geographical area of responsibility. It is the only sizeable physical science lab on base.

These chemists and physicists play a key role in the center's mission accomplishment through the use of their analytical and scientific skills. Their efforts improve production methods and save materiel and manhours, thus providing more Air Force for the dollar.

One of the new technology improvements was made with the construction of a major solar energy system to provide heated water for washing and corrosion treatment of C-130, C-141, and F-15 aircraft. This system, which cost almost $1 million, covers an acre of land with flat plate solar energy collectors using the sun rather than fuel oil to heat the water. These collectors are readily visible from Georgia Highway 247 and serve as a reminder to the public that the Air Force is turning to alternative sources of energy.

Additional management actions were also initiated to further support both local and national energy conservation goals. A Robins Energy Management Program/Energy Master Plan was instituted in the summer of 1980 to incorporate all phases of the Robins Energy Conservation Program into one quick-reference, comprehensive document to provide all organizations with a working knowledge of the energy program, policies, and goals, such as reducing operations on holiday weekends to conserve utilities, and to provide the impetus for energy studies to further improve conservation.

Computers Pervade Depot Support

The 1970s brought about monumental changes in computer systems. Large-scale, "third generation," computers were introduced such as the IBM 360, which could handle complex programs not even imagined prior to this time. The "first and second generation" computers used by AFLC, at that time the biggest computer user in the Department of Defense, were already overtaxed. However, with the tremendous cost of weapon systems in the 1970s, strict control had to be maintained to reduce the quantities of spares required, and consequently the amount of Air Force financial investment involved in spare parts.

Control Data CYBER 73 computer console

The UNIVAC 1105 and IBM 7080 were predecessors of today's third generation computers, the IBM 360 and the CYBER 73.

IBM 360 computer console

The Air Force course was already charted towards dependence on computers—the "third generation," large-scale type that processed data in billionths of seconds on micro-miniaturized monolithic circuits and stored the data on high-speed memory discs. In response to this requirement, an "Advanced Logistics System" (ALS) concept was born within AFLC to design a comprehensive, multiple-programmed system to determine materiel requirements, cut down the amount of supplies in the pipeline, and improve distribution.

ALS was to replace 92 computers and combine 257 individual data systems into one gigantic program using one data bank. This would conserve information and program storage space inside the computer and would ensure that everyone was using the same, current data.

However, the system designed turned out to be too large for existing computers, and was even about five years ahead of the state-of-the-art. Expansion of computer technology during the early and middle 1970s rapidly closed that gap to about one year but it was already too late; AFLC had promised too much to the Congress and couldn't deliver—ALS had been scrapped.

Since everyone had been pinning their hopes on the Advanced Logistics System, other computer progress within Air Force programs had come to a virtual standstill. Current systems had fallen drastically behind the state-of-the-art. Now the Air Force would have to go full speed ahead just to catch up. Some individual, "stand-alone" systems and procedures were developed to bridge the gap and avoid the complexity pitfall surrounding the ALS, and some existing procedures were revised to incorporate segments of the ALS design as updates to current systems.

An excellent example of the latter is the Transportation Automated Addressing and Labeling System (TAALS). This computer controlled system, a direct spin-off from the ALS, is installed in the shipping area to perform shipment cancellations, compute transportation charges, prepare transportation documents, accumulate production count data, perform cost comparisons, and prepare small parcel billing documents. The TAALS equipment also automatically measures and weighs boxes moving on its conveyor lines.

Another example designed to replace a portion of the ill-fated ALS almost became too complex within itself. Considerable effort on the part of its designers and implementers, however, finally stabilized the computerized Warehouse Inventory Control System (WICS) and it became a viable, exceptionally productive warehousing system, more than adequately satisfying the expectations of better utilization of warehouse space, reduction in manpower, and faster response to customer requisitions.

Not only has WR-ALC become one of the Air Force's principal users of computers, it has also become one of the largest business-type computer complexes in the South. As a result of workload assignments during the 1970s, WR-ALC now provides logistical management for more than thirty computer systems, including central logistics management of the DOD Worldwide Military Command and Control System (WWMCCS) computers, involving worldwide support for Automated Data Processing Equipment (ADPE) installed at Command and Control Centers throughout the DOD, including Air Force, Army, Navy, Marines, and Intelligence Agencies.

The Advanced Logistics System (ALS) effort did not succeed as a total system, and even though many facets of the ALS have been incorporated in other existing systems and stand-alones, a totally new effort began in the late 1970s to develop a new computer system to replace the old Stock Control and Distribution System with a modernized version, with the emphasis towards automated control and visibility. The rapid expansion of computer technology during the late 1970s and early 1980s will undoubtedly provide a greatly improved chance for success of the new Stock Control and Distribution System (SC&D).

The Advanced Airborne Command Post, the E-4, is a modified version of the Boeing 747B jet transport. Using improved communications and data processing equipment managed by WR-ALC, the E-4 provides SAC and the National Military Command authorities with increased survivability and growth capabilities in the airborne command post system.

Machine Tool Automation—Quicker Support to the Aircraft Line

Remember when machine tools were manually operated by craftsmen? The process was often tiring, slow and cumbersome; but it worked.

At Robins AFB, it worked for some thirty years. In 1972 a revolutionary concept came upon the scene—Numerical Control (NC). This is the operation of machine tools by a series of coded instructions—numbers and other symbology.

Key to the NC operation is the part programmer, who studies a workpiece drawing, prepares manuscript data, and converts it to a perforated tape which can be acted upon by the machine control unit. This program reader actually reads the coded commands for machine tools and converts coded instructions to machine-tool action. Thus, what the operator formerly did visually, the program reader does electronically with more accuracy and speed.

Numerical Control is consistent in its control of drilling, milling, boring, tapping, and other machining methods, whereas, the traditional manual machine operation depended on the machine operator's skill and knowledge, which was inconsistent.

This process has increased machine use from an average of twenty percent to a new level of seventy percent. This has been done by improving the control factor so that wasted, redundant, and nonproductive motions and waiting times are reduced or eliminated.

Eight NC machines are located in the machine shop in the Directorate of Maintenance's Industrial Products Division, Building 140, in which twenty-three persons are employed. Elsewhere in the building is the Conventional Control Machine Shop, which employs 115 persons. They operate standard lathes, mills, drills, and other machine tools. The eight NC machines require four to six times less manpower to do the job than the conventional shop and they handle at least half of the work produced in the machine shop.

In 1977, NC was followed by the Computer Numerical Control. By adding a computer to the control system, the operator has some feedback into the system by editing the program. Whereas NC operators could only do what was on the punched tape. Computer NC is more flexible than its first generation predecessor.

Computer-Aided Design/Computer-Aided Manufacturing (CAD/CAM) evolved from CNC. The principal behind CAD/CAM simply stated is: digital electronic computers, with their speed and accuracy, can help streamline design and manufacturing.

A computer graphics system was recently installed at Robins. The heart of the CAD, this computer graphics system creates, transforms, and displays pictorial and symbolic data. It serves as a high speed electronic source and replacement for the traditional drawings and documents used in manufacturing.

Improved consistency and yield of production created by CAD/CAM will, give more versatility to machine tools, relieve some vital skills shortages, reduce inventory through closer control and better scheduling, and reduce space and energy requirements.

At its ultimate level of development the computer graphics will provide WR-ALC with some completely computer-controlled processes.

Is this the dawn of Robotics at Robins? Those in the Computer Numerical Control Unit do admit that robots could be used to load and unload parts in the machines and to perform other repetitive jobs. There also is a contract with Georgia Institute of Technology to look into the application of robots at Robins. Results should be in within a year to eighteen months.

M. A. Berkner, a machinist in the Industrial Products Division, manufactures a rib cap used on a C-141 wing with hydratel numerical control equipment.

A machinist manufactures parts on numerical control equipment in support of aircraft repair.

Expanding Technology Leads to Workload Redistribution

In the 1950s and 1960s, the primary WRAMA workloads consisted of management of aircraft, aircraft spares and accessories, aircraft modifications and overhaul, and repair of aircraft components and ground support equipment. As the requirement for more sophisticated equipment and weapons generated to support the effort in Southeast Asia, the WRAMA scene began to change to more electronics and less aircraft repair, especially since most of the older aircraft were either almost out of the active inventory, had been sold to Foreign Military Sales (FMS) countries, or were in use in Southeast Asia, and were consequently being repaired over there. In the long run, it was this swing to electronics that would lead to a major change in workloads and secure the permanence of the ALC.

Under the Technology Repair Center (TRC) concept, which began in the middle 1970s, workloads were reassigned to the various Air Logistics Centers to take advantage of similar skills availability, and WR-ALC became the avionics center of the Air Force Logistics Command. It should be noted, however, that this new type workload did not seriously impair the ALC's ability to maintain overhaul and modification lines, and capabilities, for its major aircraft, the F-15, C-130 and C-141.

The TRC program was developed to consolidate similar workloads throughout the command, thereby moving the ALCs from generalization to specialization of repair and maintenance of Air Force equipment. In countering the challenge of austerity, HQ AFLC felt it should provide for workload realignment so that all the repair workloads requiring similar facilities, equipment, and skills would be located at the same ALC. Thus, each would specialize in certain repair workloads, productivity would increase, and each ALC would be able to obtain more efficient utilization of plant facilities and equipment.

Initiating this new approach to technology in maintenance, Project *Pacer Torch* was implemented in earnest in 1974 when the workloads for landing gear, weapons, hydraulics, photographics, electrical components, electro/mechanical AGE, and instruments were transferred from WR-ALC to the other four ALCs. During this same year, workloads for airborne electronics, gyros, and life support equipment were transferred to WR-ALC.

Project *Pacer Torch* was finally completed in late 1975 with over fifty individual plans involved. This effort resulted in the redistribution of more than six

Avionics Center of the Air Force

Entrance to Robins Avionics Repair Complex.

Just how did the Warner Robins Air Logistics Center become known as the "Avionics Center of the Air Force?"

Planners in the Directorates of Maintenance and Plans and Programs in the middle 1960s observed that the Warner Robins Air Materiel Area was doing a lot of its electronic work on a project basis. Even though WRAMA managed most of the electronic items, it turned projects over to San Antonio AMA, Texas, and Oklahoma City AMA, Oklahoma, for repair. Thus, WRAMA could not plan consistently on the amount of electronic workload it could handle in the maintenance shops since it was virtually all project workload.

Realizing WRAMA had the capability to expand in the avionics area, planners worked up a briefing to give the WRAMA Commander, General Beck. In it they proposed to build a facility, Building 640, the current site for the Airborne Electronics Division; and to modify Building 645, another electronics facility.

This would allow the center to handle several million manhours of work in the future. They also proposed converting manpower spaces to the electronics area and asked for additional electronics workload without added manpower.

After it was approved by General Beck, they briefed Gen. Jack Merrill, Commander of AFLC, and his staff. One of the key recommendations in the AFLC briefing was to transfer radar altimeters, radios, and electronic warfare equipment from Oklahoma City and San Antonio to WRAMA. Since WRAMA managed most of these items, collocating repair with the engineering and management was another advantage to be gained by this realignment. General Merrill was sold on the plan and specified that four hundred thousand manhours of electronics work be transferred from Oklahoma City and San Antonio to Robins. Since it had the technology base, airborne electronics work would be concentrated at WRAMA.

This realignment occurred in 1970-71. Future expansion also was assured because management made sure that the AMA was staffed to accommodate additional future workloads.

In 1973, WRAMA was named technology repair center for airborne avionics. It has since expanded to the degree that it is recognized as the number one technology repair center for the Air Force and perhaps the largest facility in the free world exclusively dedicated to avionics repair.

An electronic mechanic in the Airborne Electronics Division maintains the PAVE SPIKE pod which uses latest state-of-the-art technology.

million manhours of workload between the five ALCs. In summary, WR-ALC gained 2,021,000 manhours and lost 905,000 for a net gain of 1,116,000 manhours.

There were 875 personnel transfer offers issued relative to the gained workload assignments and 176 workers actually transferred from the other ALCs to Robins AFB with their jobs. This represented a twenty percent acceptance rate compared to a six percent rate for those moving with the lost workloads to the other ALCs. In all, Maintenance gained 510 man years of workload as a result of the TRC concept. However, even though the workload increased overall, Maintenance lost approximately three hundred employees due to Department of Defense budget cuts which occurred at this same time.

To meet the fluctuating manpower requirements during peak workload periods, the Directorate of Maintenance began, in the middle 1970s, to use Technical Services Occupation (TSO) employees. These employees were both skilled and semiskilled in various trades and crafts, and were nongovernment workers who had expressed a desire to work, when and if work became available. These on-call TSO employees were hired to work during peak workload periods and were subsequently placed in a non-duty status at the end of the peak periods. In essence, the TSOs were temporary hires. Of significant advantage to the ALC was that it could tap this reserve pool of employees without the necessity of advertising and recruiting.

Randy Mecham repairs a component of the ASQ-151 infrared system in the avionics shop.

Project Pacer Torch realigned workloads between ALCs.

The airborne electronic systems for the AWACS were assigned to WR-ALC in 1974.

Automatic Test Equipment

One reason the Warner Robins Air Logistics Center is recognized as the Avionics Center of the Air Force is due to the efficiency of its avionics testing in the Airborne Electronics Division.

The first automatic test equipment in the late 1950s was more like a circuit analyzer than an electronic tester. It was known as an analog tester.

A decade later, in the 1960s, there were semi-automatic testers. Called "general programmers," they were known as General Purpose Automatic Testing Systems (GPATS) and were used to test the complete fire control system on the aircraft. They also were analog testers. GPATS was comprised of "building blocks," each one calculated to test certain functions. GPATS could not test entire fire control systems, however, because there are some functions that can't be tested, i.e., mechanical, some electro-mechanical, optical, some inertial functions, etc. The big advantage of automatic test equipment (ATE), in addition to speed, accuracy, consistency, etc., is that it tests functions instead of things. Therefore, instead of a peculiar tester to test a peculiar thing only, ATE can test anything that performs the same function, i.e., there could be dozens of different size and nomenclatured black boxes that could be tested by the same ATE station, if the boxes all performed the same function. You only had to have test instructions (programs) for the various boxes rather than individual testers (hardware). Hence, ATE is like a player piano—it plays anything for which you have a roll, rather than a separate piano (or player, as in this instance) for each song.

Digital testers came onto the scene in the 1970s. They are computer-operated and computer-controlled test stations and are used at Robins for automatic testing. The testing is done by use of compilers, translators, digital pattern generators, and micro-processors. Compilations and tests can be performed in seconds or minutes that would require days or even weeks if done manually. Automated test equipment has provided much improvement in the rate and accuracy of fault isolation and test processes.

Automatic Digital Avionics Test Set.

The Big Picture at Robins AFB

The great majority of the public, and even many of the personnel employed at the base, are only aware of the major systems and aircraft that make the news headlines, such as the F-15 Eagle, C-130 and C-141, Strategic Air Command's B-52, and some special programs like the gunships or Pave Paws building. These are unusually large workloads affecting many manpower spaces or are unusual subjects. That's why they're in the headlines. However, very few employees or members of the general public ever become aware of just how extensive the WR-ALC management scope really is, and what the tenants' missions encompass.

WRAMA/WR-ALC managed aircraft (C-130 and HH-53) were used extensively in recovery programs supporting NASA space shots during the 1960s and 1970s.

WR-ALC's reputation for rapid response is enhanced by the numerous daily pickups (and deliveries) from the Air Freight Terminal by LOGAIR aircraft.

The ALC workload is constantly fluctuating, upwards for the most part, but makes the papers only when major systems come or go. As an example, in one year alone (1978), WR-ALC picked up workload responsibility for fourteen systems/subsystems, while losing only two. The two outgoing systems were the H-19 helicopters and the AGM-12 missiles, which were retired or deactivated. The workloads added were the following:

- The Vinson and Parkhill Tactical Secure Voice Program
- A Class IV Modification to the C-141 Radar System
- Aircraft Fire Fighting Equipment
- Procurement of the Airborne Video Tape Recorder (AVTR) System
- The AN/ALE-40 Chaff and Flare Dispenser System
- The Wide-Body High-Lift Aircraft Loaders
- The AGM-78D-2 Standard ARM Missile
- The C-141 Stretch/Aerial Refueling Modification
- A Class V Modification to the HC-130 VHF/FM Radio
- Installation of ALE-40 Chaff Dispenser on AWADS and SKE equipped C-130
- The Improved AGM-45 Shrike Missile
- The F-15 Cockpit Recording System
- A Class V Modification to the F-15 ALR Radar Warning System (AN/ALR-56)
- The AN/DLQ-3B Countermeasures System.

In 1970, in a successful effort to promote small business bids on government parts contracts, the Small Business Office assembled this display of parts, showing the prospective contractor that he may be able to produce these parts in his own factory. Use of small businesses usually lowers the cost to the government.

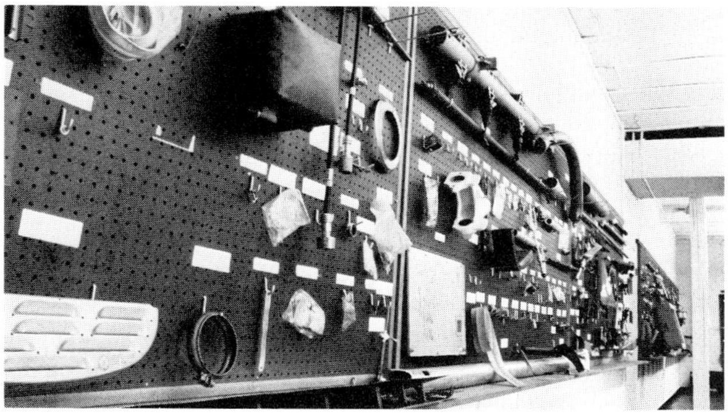

The "Stretch" YC-141B aircraft—one of our major workloads.

AC-141 tail stand in position.

At the end of the decade (1980), WR-ALC had primary management responsibility for one fighter aircraft, nine transport aircraft, five utility aircraft, four helicopters, four air-to-air missiles, three air-to-ground missiles, and three remotely piloted vehicles (small aircraft).

In addition to these aircraft and missiles the ALC

also had System Management (SM) responsibilities for the following:

The 463L Materials Handling System
The 414N Fire-Suppression and Rescue System
The Worldwide Military Command and Control System (WWMCCS)
The 437A Bare-Base System
The Air transport Airlift Control Element

WR-ALC was also Inventory Manager (IM) for the following:

Several Bomb and Gun Directing Systems
Target Acquisition (Finding) Systems
Airborne Radars
Electronic Warfare Systems
Propellers
Safety Equipment
Bearings
Vehicles
Aerospace Communications and Navigation Equipment
General Purpose Automatic Data Processing (Computer) Equipment
Airborne Satellite Communications Equipment

On the Maintenance side of the house, WR-ALC operated maintenance lines for overhaul and modification of the F-15 fighter, the C-130 and C-141 transports, and was the Technology Repair Center (TRC) for propellers, life support equipment, gyroscopes, and almost all airborne electronics.

This is one part of the tool control program to avoid foreign object damage from lost tools.

Consolidation/Containerization Point (CCP)

An additional workload added, although not in the Maintenance or Materiel Management areas, was the Consolidation/Containerization Point (CCP) in the Directorate of Distribution in late 1978.

Excessive transit time was being experienced in moving transportation priority 3 (routine) cargo to Area 2 (Germany and the United Kingdom)—in fact, eleven days more time than the average. The fact that the Air Force cargo was not containerized for shipment prior to reaching the waterport increased the problem, since that took an average of twenty-four days of processing time.

This situation caused the AFLC Commander to direct an in-depth review in November 1976. Three options were considered:

1. Use of commercial carrier (cost was too high).
2. Joint use of Army Consolidation Point at New Cumberland Army Depot (too small and too much Army control).
3. Establish an Air Force CCP at WR-ALC (WR-ALC had facility and equipment and believed it could trim thirty days off the pipeline time).

Option three was approved by the AFLC Commander in September 1978, with a one-year service test to be conducted at WR-ALC using overhires. The CCP cost $502,000 initially and it was anticipated that it would have an annual operating cost of almost $2 million. However, it would save approximately $15 million annually in port handling charges, plus a one-time pipeline saving of $2.8 to $4.6 million.

Loading a consolidation/containerization van at the CCP.

Under the CCP concept of operation, low priority surface shipments throughout the United States destined for selected bases across the Atlantic are routed to WR-ALC for consolidation/containerization into van-type containers. These vans are then moved by commercial carrier trucks to seaports in Savannah, Georgia, and Charleston, South Carolina, where they are placed on ships bound for the most convenient overseas port. At the overseas receiving point, the vans are opened and the cargo is distributed to its ultimate destinations.

This operation has been highly successful in moving cargo overseas and has reduced transit time by twenty-five days. It has been so successful that it has been expanded to include more overseas destinations in Europe and the Mediterranean, the types of cargo moved have been expanded, and another CCP has been established at the Sacramento ALC, California, to provide the same service for the Pacific area.

The C-130 Hercules

The Air Force's C-130 chapter opened in 1956 with delivery of the first Hercules from the Lockheed Aircraft Corp. By 1978, new aircraft were still being produced but the average age of active C-130s was fourteen years, and they were beginning to require close maintenance attention.

As of 1978, the C-130 fleet consisted of 720 aircraft, and even after decades in the inventory, the Air Force still had 708 C-130s flying in 1980. A Lockheed Aircraft Company analysis in 1973 had projected a structural life of 14,000 to 27,000 hours, depending upon the model. At an average of 465 flight hours per year, many C-130s would still be operational in the 1990s, and possible structural improvements could extend this beyond the year 2000.

Eight commands use nineteen configurations of the C-130 in multi-mission roles. The continued performance of the C-130 has been due in large part to the management and maintenance of the aircraft by WR-ALC. Through the years, there have been problems that threatened its excellent safety record, but corrective actions have been swift.

A throttle cable failure, in December 1978, resulted in uncontrolled full forward and reverse power and caused some serious safety incidents. WR-ALC personnel immediately took action to identify and replace all defective cables, and follow-on inspection and replacement as part of Programmed Depot Maintenance (PDM) resulted in lower downtime and more operationally available aircraft.

C-130s are versatile—but can they REALLY skip rope?

A total of 701 aircraft were inspected and 7,300 defective cables were located. All but 88 aircraft had at least one bad cable. However, since each aircraft had 45 cables, the 7,300 defective cables were not out of line with normal rejection rates.

The replacement cables were not available from stock and had to be locally manufactured by WR-ALC shops. The work was finished in a record time of three weeks, including initial purchase and delivery of the raw materials to WR-ALC. Everybody worked together and no work stoppage was experienced. Within a fifteen day period, WR-ALC made replacement cables and shipped 2,892 shipments worldwide by pilot pickup, commercial air, and *LOGAIR*.

This project was a good example in defense of organic capacity—based on the center's quick reaction.

Nose landing gear strut failures also caused several incidents in 1979 and 1980 due to stress corrosion cracks. Redesign of outer cylinders and increased frequency of inspection resolved this problem. These and other corrective actions resulted in a drop in the C-130E abort rate from 253 in July 1979 to 161 in March 1980.

In addition, the C-130 aircraft was kept up-to-date with numerous modifications to its airframe and equipment. New navigational systems, radio and radar systems were installed or planned for the future, along with modifications to replace the outer wings with new box beams, and update the engines to increase power, decrease fuel consumption, eliminate visible smoke emissions, and extend the life of the power section.

NAVSTAR Global Positioning System

One outstanding high technology program assigned to WR-ALC is the multi-billion dollar NAVSTAR Global Positioning System. This system, which is scheduled to use the space shuttle to put some of its satellities into orbit, will consist of eighteen satellites positioned around the world in three fifty-five degree inclined circular orbits, receiving information from ground stations in California, Hawaii, Alaska, and Guam, and rebroadcasting to all users of this futuristic navigational system.

Each user will be equipped with a "little black box" receiver which will use the data from the satellites to tell the user precisely where he is, anywhere on earth.

By the start of the 21st Century, it is anticipated that there will be approximately thirty thousand of the little black boxes in use and although the information will be generally available, there will be a daily code system available to lock-out unauthorized users. This system has the potential for replacing hundreds of navigational systems now in use and it can be used for accurate mapping. Its management is expected to be a major workload for WR-ALC for many years to come.

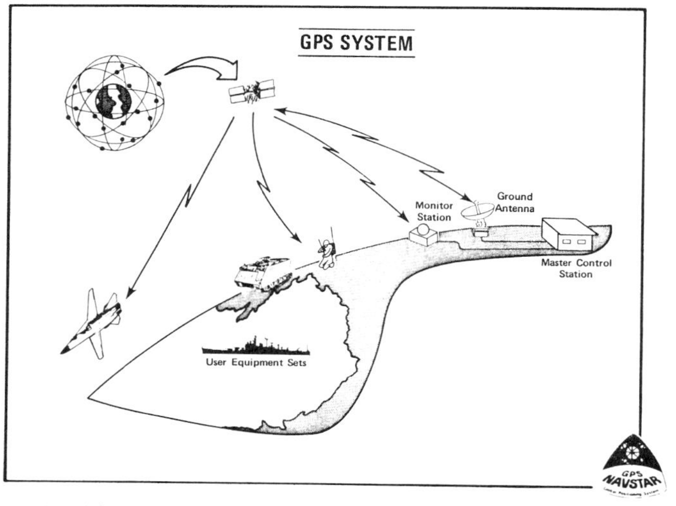

NAVSTAR Global Positioning System

Modernizing for the Future Constructing for Efficiency

The Energy Crisis, in addition to causing short-term inconveniences and minor discomforts, has had a most lasting effect on the building industry. Everybody became concerned with the construction of energy efficient buildings and what could be done to existing buildings, in the way of additional insulation, storm windows, solar heating, etc., to reduce energy consumption. The government was no exception. Without sizeable reductions in energy usage, the Government's utility bill would someday begin to resemble the national debt.

Beginning in the very early 1970s, Robins AFB embarked upon an overall building program to centralize and modernize all of the areas outside the flight line and industrial complex, i.e., housing, clubs, and service facilities. Very little had been done in this area in the way of organized improvements during the 1950s and 1960s, except for some new housing units, and even that project had been somewhat less than adequate.

The original Post Exchange as it looked in 1943.

Modern Base Exchange as it looks today.

Original Robins Field Theater.

Today's Robins AFB Theater replaced the old theater in 1973.

The original Robins Field Chapel was built in 1942 and served until 1963.

The current RAFB chapel is located across the street from the Base Exchange Complex.

This was the first Service Club at Robins Field. Note the World War II camouflage. This building was later used for the NCO Open Mess when the Service Club was moved to building 1011 (more recently the Base Library).

The Service Club while in Building 1011.

The current Service Club, the SMSgt George T. Smith, Jr. Recreation Center.

The Robins Air Force Base Golf Course is one of the best in the Middle Georgia area.

Robins AFB has several well maintained courts.

Fishing at Duck Lake.

A picnic pavillion under construction.

The new Base Youth Center, east of the NCO Club.

One major construction, beginning in 1970, was a new Base Exchange Shopping Complex designed to include the Base Exchange, a new Base Theater, a bank annex and a Post Office Branch. This complex is across the street from the Base Chapel, adjacent to the bowling alley, and across the street from the Airmen's Club.

This complex now complements, in a semi-centralized location (all within two or three blocks), the Base Commissary, the NCO Open Mess, the Base Service Station and the Base Hospital. Additionally, in 1982, a new building was constructed adjacent to the shopping complex to house the Personnel Services Center and the Robins Resident Center (RRC), finally moving that educational facility out of the old, World War II dormitory area within the city of Warner Robins.

All of this is now within easy access of the on-base housing units, including two hundred new units built in 1972. It is also readily accessible to off-base personnel through the new South Access Road, which opened in 1976 and joins Russell Parkway on the South Side of Warner Robins.

This construction program effectively divided the base into three major use, or activity, zones—the flight line, including the Strategic Air Command (SAC) Area and the WR-ALC Maintenance Shops in the hangars and adjoining buildings; the industrial/administrative complex containing offices, warehouses, and avionics shops; and the residential area on the south end. The residential area also includes excellent recreational facilities, such as a golf course, four swimming pools, a gymnasium, a lake and nature area, and many athletic playing fields and tennis courts.

This division has, to some extent, succeeded in isolating traffic within zones, and has resulted not only in added convenience but reduced automobile fuel consumption as well.

In addition to the construction in the residential service areas to improve on-base facilities for resident military personnel and dependents, considerable construction also occurred in the industrial area at Robins Air Force Base throughout the 1970s. Both total facility construction and internal remodeling were accomplished to improve mission capabilities in support of the changing workload. Much of the internal remodeling was primarily concerned with advancing technology. Construction and remodeling projects included the following:

The Early 1970s

New Engineering Test Facility in Building 226

New Flight Line Control Tower
Maintenance Industrial Improvements
New Plating and Metal Finishing Shop
Installation of a mammoth Autoclave
New Precision Measurement and Evaluation Laboratory (PMEL) in Building 162
New passenger lounge in Flight Operations Area in Building 111 to serve passengers awaiting military flights

The Middle 1970s:

Distribution Logistical Materials Processing Facility Building 376 including Shipping, Receiving, and DS Administrative Offices
Maintenance Improvements:
- Electrostatic Airless Painting System
- Solar Purge Fluid Reclamation System
- Phosphoric Acid Anodizing Process
- Electromatic II Force Calibration System

The Late 1970s:

New Automated Avionics Warehouse, Building 641
New Electronic Warfare Building
New Air Freight Terminal

The WR-ALC Engineering Test Facility (Building 226) was completed in 1974, the first segment of a growing complex.

The Precision Measurement and Evaluation Laboratory (PMEL) in Building 162.

The original Robins Field Control Tower, located on top of Building 110.

The new Flightline Control Tower is seen to the right of the C-141 tail sections.

The Automated Avionics Warehouse in Building 641. This warehouse contains the Warehouse Inventory Control System (WICS).

Interior of the giant autoclave unit installed in WRAMA maintenance facility in 1970 at a cost of $520,000, the first of its kind in AFLC. It provides the capability to cure any bonded component used on the C-130, C-141, and C-5 aircraft.

Distribution Logistical Materials Processing Facility, Building 376 (large building in right foreground). It contains shipping and receiving departments plus administrative offices.

Processing and hazardous storage areas of the Air Freight Terminal.

The first Air Freight Terminal at Robins Field—it used unsophisticated equipment for loading and offloading cargo but the spirit and cooperation of the people overcame obstacles caused by the lack of modern materials handling equipment. This first terminal was operated as a Detachment of the 39th Air Freight Wing. It was later taken over by the Military Air Transport Service (MATS) and operated by them prior to being turned over to WRAMA to operate.

The new Air Freight Terminal, completed in 1980, is one of the most modern terminal facilities in AFLC. It contains 60,000 square feet of processing area, 9,000 square feet of administrative space, and features a LOGAIR Real-Time Terminal Computer System composed of the Honeywell H700 computer and six on-line remotes connected to the Burroughs B4700 Computer. This system enables the terminal to handle up to nine million pounds of cargo monthly, processing thirteen LOGAIR and three military flights per day.

Storage pallet racks in the Air Freight Terminal.

Security around the flightline area was also increased in the late 1970s. A police walking patrol was added in 1978 to check on hangars and ramps from six at night to six in the morning on weekdays and around-the-clock on weekends and holidays. This was in addition to two existing, twenty-four hour motorized police patrols in the area.

Although this action was taken following an incident of damage to a SAC alert aircraft, the ALC commander was more concerned about the size of the police force in the area when balanced against Presidential visits, increasing F-15 traffic, the SAC weapons storage, and small arms storage in SAC and in transit through the Air Terminal.

Plans are also underway to enclose the South entrance to the flightline with a chain link fence and civilian guards in the very near future.

Area Defense Counsel

The office of the Area Defense Counsel was established in 1974 following the recommendation of a task force composed of the Judge Advocates General of the Air Force, Army, and Navy, at the direction of the Secretary of Defense. It was intended to enhance the independence and professionalism of legal defense services which had previously been rendered by subordinates of the local Staff Judge Advocate. It is now a subdivision of the Department of the Judge Advocate General, and is an independent tenant at every base where it exists.

Defense functions include counsel duties in courts-martial and administrative boards; nonjudicial punishment counselling; military legal advisor duties in foreign countries, and other adverse actions in which counsel for an individual is required or authorized; and counsel for Air Force Reserve (AFRES) discharge board actions. The Area Defense Counsel is expected and required, whenever possible, to perform all legal defense functions at the base where assigned.

Mechanization in Distribution

When the Southeast Air Depot was activated in November 1941 at Wellston, Georgia, the first Supply Division, under Lt. George E. Kegin, was established at the Atlantic Ice and Coal Company warehouse on Waterville Road in Macon. Its mission was to assist the construction contractors, expedite supplies, and rush completion of the supply buildings at Wellston and, except for what we would consider antiquated forklifts, all physical movement of materials within storage was done by hand.

Little improvement was made in handling of materials until the late 1950s when two huge warehouses were built and a giant mechanized materials handling system was set into operation. Within the two new warehouses, which covered twenty-seven acres, a one and one-half mile long towveyor system was built to move materiel to and from the shipping and receiving units located across the street in two other buildings, connected to the new warehouses by a tunnel. The towveyor system, which was a motorized drag chain moving in a channel in the concrete floor, was powered by eight electric motors and was capable of pulling 1,500 warehouse dollies, each loaded with 400 pounds of materiel, through the warehouses at 4 mph. Its main benefit was that it freed personnel to do their warehouse functions rather than having to push or pull the dollies to their destinations.

Although there were minor problems with the system, such as jamming from trash that had been swept into the chain mechanism or from dollies colliding with carelessly placed crates and trailers, it remained in use well into the 1960s.

Mechanization in the 1950s: the Distribution Towveyor System. Note the dolly following the slot in the floor.

Central Receiving and the packing and shipping areas were also mechanized, as part of this overall materials handling system, with numerous roller and belt conveyors, including remotely operated kickers and line-selection devices for sorting and routing materiel within the workcenters.

In the late 1960s, it became readily apparent that, with the oncoming need for austere manning, items would have to be stored and moved more efficiently. In 1969, a small item warehouse was established in Building 385, the largest of the two recently built ware-

Mechanization in the 1960s: Central Receiving and Surface Terminal.

houses, and the main design ingredient was further mechanization featuring a monorail conveyor system with photoelectric coded tote-trays to direct materiel to and from the proper locations, along with a magnetic in-floor towcart system covering over sixteen hundred feet on the ground floor level. This facility, with its specially designed features for easier materials handling, reduced response time for issue and receipt transactions for small items, required less manpower, and improved small item handling overall.

Two major construction projects were completed during the 1970s that involved even greater mechanization; a Logistical Materials Processing Facility and an Automated Avionics Warehouse, both highly automated and containing maintenance shops expressly for the repair and upkeep of the conveyor systems and other materials handling equipment therein.

The Logistical Materials Processing Facility, Building 376, along with the Directorate of Distribution's administrative offices, houses freight receiving, preservation and packing, and shipping departments. All are highly mechanized with belt and roller conveyors, an overhead towveyor sortation system, and computerized weighing and labeling equipment. This facility is connected via an underground tunnel to the warehouse containing small items, and it will someday be connected to the Avionics Warehouse and the Air Freight Terminal by DARTS, an automated materials transportation system using driverless, programmed, wire-guided tugs to pull trailers loaded with materiel.

Mechanized conveyors in the 1960s.

Mechanization in the 1970s: The surface Terminal in the Distribution Logistical Materials Processing Facility, Building 376.

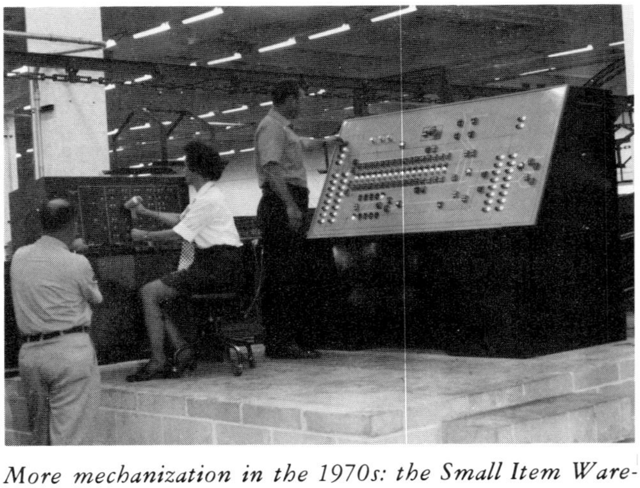

More mechanization in the 1970s: the Small Item Warehouse automated monorail conveyor system control panel.

The Avionics Warehouse, completed in 1977 adjacent to the Maintenance Avionics Repair Facility in Building 640, provides an automated storage and materials processing capability to speed the movement of the thousands of items stored therein. The most important feature of this automated warehouse is the computer-controlled Warehouse Inventory Control System (WICS). The WICS is a computerized materials handling system, utilizing Remote Vehicle Processors (RVPs), somewhat similar to modern fork lifts but containing on-board computers, to store and retrieve materiel. The RVPs are capable of taking orders and selecting items from storage locations, and were designed to speed the issue and movement of the thousands of items stored in the Distribution Avionics Warehouse.

The heart of the WICS consists of four Data General computers tied together by a communications adapter that allows them to function as one unit in the role of mechanized materials handling equipment controller.

Overhead monorail conveyor system and magnetically controlled dollies used in the Small Item Warehouse.

A WICS Remote Vehicle Processor (RVP) at ground level.

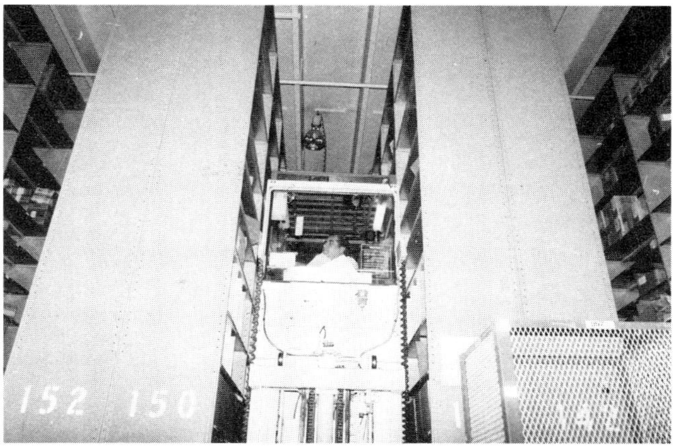

A WICS RVP at work storing materiel in the high bay storage area.

Although the system is called a Warehouse Inventory Control System, it provides more than just inventory information. In addition to receiving and storing information concerning items stored in the Avionics Warehouse, it interfaces with the Stock Control and Distribution System (DO33) and the Shipment Document Release System (D009) to select items for off-base shipment and to provide spare parts to the adjoining maintenance shops.

All items are placed in and retrieved from storage locations by personnel riding on the RVPs, containing on-board computers which communicate with the central WICS computers. These RVPs, with computer guidance, follow wires embedded in the concrete floor to move to the selected storage locations. Additionally, the RVP Computers receive and transmit vehicle locations and work process information to and from the central WICS Computers to keep a constant update of materiel stored in the warehouse and handle traffic control of the RVPs.

There is also a conveyor system to move materiel to and from various points in the warehouse, and to and from the packaging workcenter within the building.

Also, the Small Item Transfer (SIT) System, a pneumatic tube system, connects the three buildings (640, 641 and 645) at strategic locations. This permits a thirty-minute response time for small parts going to the avionics repair shops.

This automated storage system provides better utilization of warehouse space, reduction in manpower, and faster response to customer requisitions, thereby increasing productivity.

Additional mechanization is planned for the Directorate of Distribution in the near future, in the form of another warehouse system, currently under development, similar to WICS, side-loading and narrow-aisle forklifts to reduce wasted aisle space, and the previously mentioned Distribution Automated Rapid Transportation System (DARTS), which will provide programmed, wire-guided tugs to pull trailers between buildings.

A recently added automation feature in Central Receiving is LOGMARS (Logistical Marking System), which uses a "magic wand," actually a light-sensor wand, to read bar-coding applied to items being received, and then inputs the data to match prepositioned records in the computer, reducing human effort considerably and speeding up the receipt process by about four hundred percent.

Meeting the Needs of Tomorrow Logistics Management and Readiness

Increased concern developed during the later 1970s over preparation for operational readiness in a crisis situation or wartime scenario. Over the years, organizations had become complacent and tended to exist only because they were there, and the growing concern was whether the command was in a proper state of readiness. Deficiencies discovered after the fact would come too late to benefit the command.

The term "readiness" viewed from an Air Logistics Center's standpoint involves, among other things, logistical support ability of assigned weapons systems and managed items, capability to support wartime conditions from manpower, skills, facilities, and equipment aspects, and capabilities as an air cargo airdrome, including mobility processing.

The key question in all of this concern, however, is the measurement of readiness. There are no specific reports or criteria which, from an ALC perspective, measure the total readiness of an ALC or the logistic supportability of assigned system/item management missions.

The establishment of a Logistics Readiness Center (LRC) in the Command Section at WR-ALC and the assignment of a project officer in the Plans and Programs Directorate (XR) as a focal point for readiness planning seeks a resolution of this problem.

5th Combat Communications Group

The 5th Combat Communications Group has been located at Robins since July 1964. Throughout the last two decades the group has provided emergency communications and air traffic control around the globe as well as supporting major Joint Chiefs of Staff exercises.

During exercise Proud Phantom in the summer of 1980, members of the Fifth were among the first American military forces to set foot on Egyptian soil since World War II. They provided voice and data communications via satellite with the United States and Europe within twenty-four hours after their arrival. Also, they successfully demonstrated the first application of long distance remoting of a supply computer.

Repair Teams

Robins 2955th personnel repair a fire damaged F-15 in Bitburg, Germany. Left to right: *SSgt. Roy Sims, SSgt. Larry Mason, SSgt. Jim Sager, and SSgt. Dean Dunn.*

Ever since the Vietnam war, the Warner Robins Air Logistics Center has extensively used augmentation teams—maintenance, transportation, and supply teams—who literally "take the depot to the user."

In 1968, the 2955th Combat Logistics Support Squadron (CLSS) was formed in conjunction with the Rapid Area Maintenance (RAM) and Rapid Area Supply Support (RASS) Teams. Its military teams were deployed to provide depot repair and modification work in the field.

Since then, the 2955th CLSS has increased the scope of its mission to provide rapid aircraft battle damage repair and combat packaging and supply operations.

The Transportation and Maintenance Branches of the 2955th CLSS deploy combat distribution teams, rapid aircraft battle damage augmentation repair teams, combat aircraft repair teams and rapid air maintenance teams to handle work in the field.

These 2955th CLSS military teams often work with civilian maintenance technicians at the depot while on temporary duty (TDY) or on the flight-line on the F-15 at Robins. Occasionally, they will be accompanied on TDY trips by civilian maintenance workers.

This is but another example of the military-civilian teamwork which has become a trademark for the Warner Robins Air Logistics Center.

Electronic Warfare Reprogramming

Military victory or defeat in any future conflict will be highly dependent upon the technological state of military electronics. Although electronic countermeasures played a significant role in World War II with the use of radar, it was not until the Southeast Asia Conflict that today's electronic warfare (EW) systems began to take shape.

Initially, WRAMA became involved in electronic warfare reprogramming in the late 1960s, when a team of WRAMA engineers was assigned the task of making several improvements to the EW systems preceding the ALR-46 Radar Warning Receiver System. The quick reaction capabilities which were developed, to counter and penetrate the elaborate USSR—North Vietnamese electronic defense system over North Vietnam, were the evolutionary beginning of the most extensive EW engineering technology in the Air Force.

During the Middle East conflict in October 1973, Eqyptian ground forces, using surface-to-air missiles (SAMs) supplied by the USSR, were successful in shooting down some one hundred Israeli aircraft. Engineers from WRAMA were called upon to immediately reprogram the Israeli EW systems to counter the SAM challenge. Fortunately, a captured SAM was available for our engineers to examine the Soviets' technology. As a result, the hurried reprogramming measures were successful and the war ended in a cease-fire.

Although the lessons learned from the Middle East war were most significant, the United States still needed to be better prepared to meet future challenges, since the complexity of reprogramming varies considerably depending upon what the system is to accomplish. Some reprogramming, involving only minor data changes, can be done in a few hours while extensive modifications sometimes require several months of development and reprogramming.

The requirement for the Air Force to control the EW reprogramming followed both from the operational requirement for quick turnaround of changes and from the nature of the systems themselves. The Air Force was investing large sums of money to develop and acquire flexible reprogramming systems. The ability to use this flexibility was the basis for the Electronic Warfare Avionics Integration Support Facility (EWAISF) concept, that collection of equipment, personnel, and operating concepts for redesign and reprogramming of EW systems. Tactical tradeoffs were involved in the software development of most

Into the Wild Blue Electronically

The reader should know by now that the Warner Robins Air Logistics Center calls itself the "Avionics Center of the Air Force." He probably has only a hazy idea of what that means or the revolution which avionics has wrought in aerial warfare. The following article reprinted by permission from the June 21, 1982 issue of *Time* magazine provides a vivid picture. The F-15, the Sidewinder, and Shrike missiles, as well as the electronic countermeasures and head-up display mentioned are managed by the Warner Robins Air Logistics Center.

Israeli F-16: Star Wars electronics, laser beams and a supersensitive Sidewinder

Into the Wild Blue Electronically

The successful Israeli air strikes on Syrian SA-6 mobile missiles in Lebanon's Bekaa Valley demonstrated that modern air war is as much a matter of computer array *vs.* computer array as man against man.

The Israeli attacking force, more than 90 strong, featured U.S.-built F-15s and F-16s, two of the finest fighters in the world, flying electronic marvels. Both planes are equipped with a system called "head-up display," or HUD; projected on the pilot's windshield, in phosphorescent green and orange, is a mass of essential data. An F-15 pilot flashing over the Bekaa could have watched the plotted positions of four separate enemy aircraft and also have been alerted by a flashing light and beeping in his headset if an SA-6 radar locked on to his jet. The F-15's computerized electronic countermeasure (ECM) equipment would have taken over, perhaps electronically "disguising" the F-15 to Syrian radar, then determining which enemy threat was the most immediate. If a jet-propelled SA-6 were fired, traveling toward the Israeli plane at 2,000 m.p.h., the jet's ECM would have singled it out for intense electronic jamming, trying to overcome the SAM's own antijamming system to send the missile veering off course. If one of the Syrian MiG-21s and MiG-23s had fired a Soviet Atoll missile at an Israeli attacker, the same ECM beaming might have "spoofed" the weapon and forced it harmlessly off course.

When stalking Syrian fighters, an F-16 pilot, again using the HUD windshield images, had only to position a computer-generated "tracerline" over the MiG and fire his machine-cannon for one second; 100 20-mm rounds would have streaked toward the enemy jet. However, many of the downed MiGs were apparently hit by the latest model of the Sidewinder missile, which is being used effectively by British Sea Harriers in the Falkland Islands. The weapon adjusts its aim in flight with an extremely sensitive infrared homing system that guides the warhead toward the enemy jet at 1,650 m.p.h., faster than 95% of the planes in the Syrian air force. The new Sidewinders are so sensitive that they can lock in on the heat created by air friction on a jet's surface. The Israeli Sidewinders, like most of the electronic equipment on the aircraft, are not standard U.S. issue: they have been adapted and improved in the light of experience gained by Israel's constant sorties over Lebanon since 1976.

Israel's air-to-ground missiles are also formidable examples of *Star Wars* electronics. The U.S.-built Shrike missiles are designed specifically to knock out ground-based antiaircraft batteries such as the SA-6s. From as far away as 25 miles, the Shrikes' radar-seeking device can be tuned precisely to the SAM's frequency—probably recorded by Israeli drone planes flown over the area before the strikes. The Shrike missiles could then home on the target, effectively turning the enemy missile control system against itself.

Some of the SAM sites were probably wiped out by Maverick air-to-surface missiles, or "smart" bombs, some of which are guided along a pinpoint beam of laser light to within a few feet of the target. They are comparatively slow, but still accurate.

Syria's MiG-21s, MiG-23s and MiG-25s, although relatively stripped-down export models, do have ECM gadgets, early-warning systems and air-to-air missiles. But their electronics are not as precise and powerful as the U.S.-Israeli counterparts. Israel also had available an even more powerful electronic back-up system: four U.S.-built E-2C Hawkeye surveillance planes, each able to track 250 enemy aircraft up to 300 miles away. In addition, Israel's pilots are among the best in the world. No fighter pilots have more intensive training—in ground practice with computerized simulators, in mock air battles against captured MiGs and in actual combat. If the Israeli claims that they destroyed 79 Syrian aircraft in last week's battles while losing only one themselves are to be believed, the training has paid off.

EW systems and it was necessary that the Air Force control those tradeoffs rather than depend on full contractor support.

When the requirement came to determine the feasibility of developing and installing equipment to provide software control of the EW systems, the WR-ALC engineers were selected because of their success with earlier quick reaction capabilities. When the WR-ALC engineers proved the feasibility of software control, their engineering capability to accomplish further developments became recognized.

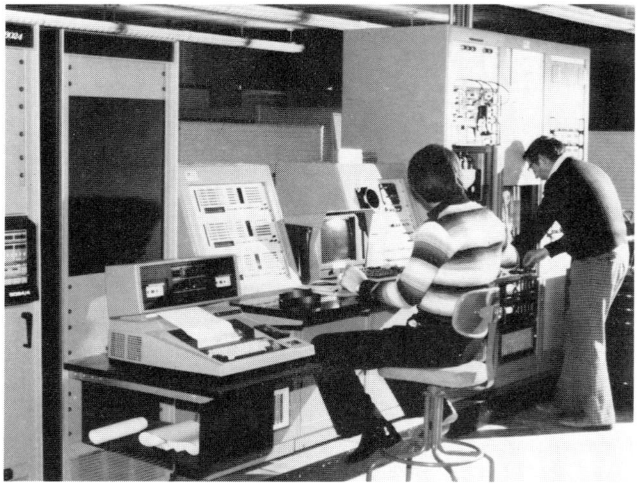

Software Engineers in the F-15 Tactical Electronic Warfare Systems Unit.

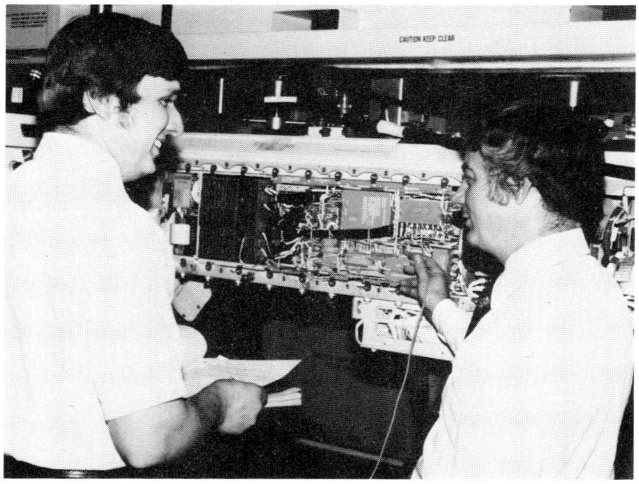

ALQ-131 Electronic Countermeasure Pod.

WR-ALC recognized that the increased sophistication of EW systems would bring about increased equipment costs and that, as a result, available resources would have to be used efficiently. The center adopted the position that centralization of all electronic warfare reprogramming at WR-ALC was both efficient and logical, and would assure maximum commonality between systems along with minimum duplication of both technical and support equipment. Collocation with the item manager, engineering, procurement, and depot repair activities would assure the most supportable response to operational requirements.

Finally, in May 1975, Air Force Headquarters issued a directive which established WR-ALC as the reprogramming center for all Air Force electronic warfare systems and authorized funds for accommodating the workload. Although not all EW activities could be accomplished at the WR-ALC facility, the center could still control those tasks placed on contract.

By the end of 1977, WR-ALC was supporting several versions of the ALR-46 Radar Warning System, used extensively by SAC and TAC, and had developed improvements to the basic system to include response to many enemy threats not previously identified. The center was also involved in the integration role for the Tactical Integrated Electronic Warfare System (*Compass Tie*) which coupled jammer-power management to the radar warning receiver to provide a response capability against continuous wave radars, ultimately resulting in equipment for use on the F-16 aircraft.

WR-ALC also developed in-house support capability for the APR-38 Wild Weasel, the F-15 Tactical Electronic Warfare System, the ALR-62 Radar Warning Receiver, the ALQ-131 Jamming Pod, the EF-111A aircraft Electronic Warning System, and several other minor EW systems. Altogether, by 1981, WR-ALC had over fifty million dollars worth of equipment and development costs invested in its support facility.

Reprogramming tasks have constantly changed because of new technological changes introduced into systems under development. The minicomputer of the early 1970s has been complimented by multiple microprocessors throughout the EW systems. New memory types introduced plus reduced packaging techniques have made tasks feasible that were not previously considered possible. Programs within these systems have grown from a few thousand program instructions to tens of thousands, and will surpass one hundred thousand on some new systems. The WR-ALC technical base has had to stay abreast of these advances through continual education and infusion of knowledge, both from newly hired personnel fresh out of college and from contractor technical exchanges.

As change requests have become more widespread, mostly because of threat changes recognized intelligence and operational units, across several electronic warfare systems, the work impact on the center has worsened. Since the frequency of requests has not declined, and is not expected to do so in the future, WR-ALC has had to adjust its capability upwards so as not to fall further behind. This is being accomplished by obtaining needed resources and facilities, by adopting time-saving efficient procedures and concepts, and through increased management control of work activities and manpower.

Production Base Expansion

In the midst of our concerns over readiness, industry, on the whole, was becoming limited in its capacity to rapidly expand its production base. This was, and is, an international phenomenon brought about by economic conditions. Economic forces, including tax structures, have inhibited industrial modernization and expansion, and have discouraged industries from stockpiling raw materials and finished goods. The trend in industry is now towards procuring raw stock after receipt of an order. The lack of materials and the lack of production capacity has added excessive lead-times. Additionally, WR-ALC has encountered situations where private industry has bought out the entire capacity of certain U. S. manufacturers, forcing the Government to look for new suppliers, both domestic and foreign.

To be prepared for contingencies and their associated production surges, workloading decisions and actions must be based upon requirements to be placed upon the command in time of war, not peace. Although surge computations in the past have been based on the gross flying hour program, some concern is now being expressed that it may be an unreliable indicator of workload, and may not provide the command with the necessary data to determine valid surge requirements.

The visibility of surge requirements through stock number level is needed, and that visibility needs to be built into the data systems used by the system managers and item managers. Additionally, test equipment must be bought with production surge in mind rather than maximum peacetime utilization. WR-ALC has consistently urged that future efforts be directed towards providing surge capability. Efforts are continuing although no easy solutions have been identified.

However, planning for improved production support is continuing in other areas. Distribution, through a DOD program called LOGMARS (Logistics Applications of Automated Marking and Reading Symbols) which uses bar coding similar to that used on grocery items, will soon begin to electronically sense and input receipt data to match-up with prepositioned computer records. This electronic receipt processing will increase receipt processing speed by almost four hundred percent, and will reduce posting costs by about thirty-eight percent. In effect, Distribution will be able to expand its receiving department by about four times with no additional requirement for floor space.

Education and Training

Improvement of the workforce through the endless process of education and training has historically remained among the top priorities at Robins Air Force Base. Although in the beginning years the problem was shortage of technical skills in the local area, by the 1960s it had become apparent that many who were hired in the earlier years simply could not keep pace with the advancing technology. A significant portion of the workforce lacked the formal education normally associated with their position requirements.

The establishment of the Robins Resident Center (RRC), in 1970 was a large step toward closing the education-technology gap. Along with the continued emphasis placed on attending the RRC to improve the overall educational level during the 1970s, several special categories of jobs were created to permit individuals to work and advance their education concurrently. Among these were the Vocational-Technology (Vo-Tech), Co-op, and Growing Engineers' programs.

One of the co-op students, Rhett Glover from Middle Georgia College, operating a computer terminal in the F-15 Avionics Integrated Support Facility.

The RRC has surpassed the most optimistic predictions of state educators and base officials both in enrollment and in enthusiasm generated. The RRC now also offers several graduate program options in addition to its undergraduate program.

The improving educational levels now foster additional, and more technical, in-house training to adapt to the new skills and technology requirements.

AFLC NCO Academy

General Bryce Poe, AFLC Commander, addressed a 1981 NCO Academy graduating class on the occassion of the school's twenty-fifth anniversary. Left to right: *Mrs. Bryce Poe, Mrs. James Jessee, TSgt. James H. Jessee (Honor Graduate) and General Poe.*

The AFLC Noncommissioned Officer Academy was initially established in 1956 at McClellan Air Force Base, California, and the first class of seventeen senior NCOs was graduated on October 19 of that year.

Patterning its objectives after those of the first USAF accredited academy established in 1953 at West Drayton, England, the AFLC NCO Academy received accreditation in 1957. As a program of professional military education, the academy mission is to ensure that select NCOs are able to fulfill their proper roles in the United States Air Force. Curriculum activities are designed to prepare the student for more advanced leadership and management positions, and to ensure they are aware of their responsibilities as noncommissioned officers.

In the summer of 1970, the AFLC NCO Academy completed 14 years of activity at McClellan Air Force Base, having graduated 3,143 students. In August 1970, the AFLC NCO Academy was relocated to Robins Air Force Base, Georgia, and was consolidated with the Command NCO Leadership School already at Robins.

Redesignated the "Air Force Logistics Command Noncommissioned Officer Academy/Leadership School," operations were resumed in the new location in February 1971. A six-week course is now conducted alternately with leadership school activities and an academy class is graduated each quarter.

The Society of Logistics Engineers (SOLE)

The Society of Logistics Engineers is a nonprofit professional association for individual members, devoted to scientific, educational, and literary endeavors to enhance the art of logistics technology, education, and management.

Logistics is a combination of many functions none of which, alone, is logistics—traffic management, procurement, inventory control, maintenance, computer science, engineering, and others. To be logistics, many functions must be organized, managed, and performed as integrated systems and subsystems. When a manager, engineer, or educator recognizes the technical and economic role of the many functions in these logistics systems, and understands their interrelationships, then he is a logistician.

Dr. Werner Von Braun, the late chief of NASA, is generally regarded as being the father of SOLE, the only society dedicated to all the functions involved in the design, acquisition, distribution, maintenance, and recycling or disposal of all resources across the entire logistics spectrum.

Chapter Seven of District Three, officially established June 28, 1971 at Robins Air Force Base, has always been awards conscious and has actively presented and received awards. The chapter annually recognizes outstanding logistics students at the Robins Resident Center and provides certificates and mementos attesting to the students' achievements. Additionally, the Robins Chapter honors a member each year for outstanding contribution to logistics and to the chapter.

The Robins Chapter meets monthly at the Officers' Club and usually is addressed by an executive from the Department of Defense, Industry, or SOLE International. There are no grade or rank requirements, and the chapter is open to anyone seeking to enhance a career in logistics.

Sharing—Community Involvement

One of the Air Logistics Center's and Robins Air Force Base's main sustaining forces is their involvement in, and support of, the local communities. Under special exemption from the Federal Government, civilian employees are permitted to participate in nonpartisan politics in several of the local communities, and can serve in part-time positions, such as city council members, within those local governments. In one small, local community, one base employee was the mayor (part-time) for several terms. This involvement is only logical, however, since an unusually large proportion of the population is employed at Robins Air Force Base.

It follows then, that base employees are present in almost all neighborhoods, to varying degrees, and have a voice in most neighborhood matters, whether the concern is about school zoning, land-use, street resurfacing, or even garbage collection. Naturally,

Robins AFB has long considered that it has the best community support in the Air Force. Each commander in the last decades has established an advisory group of area citizens who maintained this closer working relationship. Pictured are members of the Citizens Advisory Board of 1973 attending a meeting with Maj. Gen. Robert E. Hails. From left to right: Dr. Dan Callahan, Jeff Pierce, Emmett Barnes, Tom Green (deceased), General Hails, Foy Evans, Bob Young, Walt Whiting, and Bill Graham.

reciprocal concern grows out of this presence; the community has a vested interest in its residents' employment by the base, and in current and future plans for base workload, employment, construction, and mission. In effect, the relationship is a well-traveled, two-way street. Local citizens and merchants support the base and its activities, and base employeees in turn spend a large share of their salaries in the local communities. Both want to see the other prosper.

Several special projects have also been fostered by the base and its personnel to assist the local communities and to help build goodwill. Along with official support in the form of supplies and loaned equipment, Robins personnel, both military and civilian, freely give of their off duty time to support activities such as the Special Olympics (track and field games for handicapped young people) and the Happy Hour School for Exceptional Children (base personnel played a large role in its beginning and expansion), and the base has an official hire-the-handicapped program which is mutually beneficial. These handicapped employees have proved to be exceptionally good workers.

Numerous other activities are joint efforts and joint participation also, such as concerts by the Air Force Reserve (AFRES) Band and other similar events sponsored by the Air Force Association. In October 1971, the Air Force Association sponsored several nationally famous country-western entertainers in a five-hour entertainment special in the Maintenance paint hangar. The show headed by Junior Samples from the *Hee-Haw* TV show, was staged to raise funds for the new Air Force Museum at Dayton, Ohio.

Additionally, since Robins Air Force Base does not have its own junior and senior high schools, students in those grades attend local area junior and senior high schools, and many military parents serve on parent-teacher organizations, boards and various other committees.

On a more official level are joint ventures entered into by the base and city/county governments, such as land acquisition and transfer for beneficial use by all parties concerned. In fact, in the earlier years of the base, the military installation provided water and sewerage services to much of the city of Warner Robins,

The Special Olympics is generously supported by base organizations and personnel.

Robins Air Force Base personnel were instrumental in establishment of the Happy Hour School for Exceptional Children and the Happy Hour School Workshop.

Junior Samples of the Hee-Haw *televison show performed with several other country-western performers in an extravaganza sponsored by the Air Force Association to benefit the Air Force Museum in Dayton, Ohio. Betty Clark and Mary Howe are pictured with the star.*

Robins Air Force Base provided fire protection to Warner Robins homes and businesses until this fire station was built on Myrtle Street in 1959.

City and county governments swap lands with General Services Administration to help future growth at Robins AFB. Dominick Molinari, General Services Administration representative (left) confers with Maj. Gen. Robert E. Hails, WRAMA Commander, Warner Robins Major Ed Bryant and Houston County Commission Chairman Alton Tucker during deed exchange ceremonies at Robins AFB. The formal deed exchange concluded negotiations which added 842 acres to the base, for which Warner Robins and Houston County recieved 92 acres of downtown Warner Robins property. Maj. Gen. A. J. Beck, USAF retired, former WRAMA Commander who was the originator of the land exchange idea, attended the ceremony.

Getting ready for the annual charity drive—the Combined Federal Campaign.

and continued to provide fire protection until the late 1950s, when the city built its own fire station on Myrtle Street.

In an effort to further the overall preparation for future expansion, the base worked out a land swap with the City of Warner Robins and Houston County, completed in 1973. In return for ninety-two acres of choice, commercial property on Watson Boulevard and on North Davis Drive in the City of Warner Robins and the land lying between the railroad and Georgia Highway 247 at the southern edge of the Base, Robins Air Force Base received 842 acres of farmland adjacent to the south end of the base, plus $30,000 in cash to the U.S. Treasury. The consummation of this land exchange followed four years of intense efforts by the Mayor and Council of Warner Robins and the Commissioners of Houston County, efforts which were sustained through many changes of elected officials without dissent.

Chiefs' Aid Association

The Chiefs' Aid Association was formed in 1978 to aid enlisted personnel and other service connected persons in the Middle Georgia area who need financial assistance. This group has distributed over $3,000 to persons requiring immediate financial aid and yearly provides unsolicited relief to several families during the Christmas Season.

Operating funds for the Association are obtained through donations from the wives clubs and other individuals, special offerings from the Base Chapel congregation, annual dues of the association members, and from proceeds of the Annual Christmas Tree Sale.

This group is composed entirely of Chief Master Sergeants and Chief Master Sergeant Selectees, both active and reserve, assigned to Robins Air Force Base, and it meets on the second Tuesday of each month from September through May, at the NCO Open Mess.

Chiefs' Aid Association getting ready for the Christmas tree sale.

Following the land transfer, the Robins Federal Credit Union bought much of the land on Watson Boulevard from the city and county and built a four-story building to house their main offices and to provide offices for rent to other businesses. Although a building materials store, three fast food establishments, two used car lots and a convenience store have located on the North Davis Drive property, much of it remains available for development. The Richard B. Russell Parkway construction required much of the property between the highway and the railroad.

And of course there is the annual Combined Federal Campaign which means so much to the continued operation of charitable organizations both on the local and national levels. As an example of base involvement, Robins Air Force Base employees in 1981 contributed $691,867, with $328,862 of that total being distributed to the United Way Organizations in eight Middle Georgia counties, primarily Houston and Bibb Counties.

Distinguished Visitors to Robins Air Force Base

War heroes, high government officials, and top entertainers have visited Robins during its memorable forty-year history. Included on this list are such famous persons as comedian Bob Hope, World War I flying hero Eddie Rickenbacker, the legendary air pioneer Claire L. Chennault, and Nationalist China's first lady Madame Chiang Kai-Shek. In the 1964 presidential election campaign, both President Lyndon B. Johnson and GOP candidate Barry M. Goldwater visited Robins AFB. In November 1973, President Richard M. Nixon visited Robins on his way to attend Mr. Carl Vinson's ninetieth birthday celebration and the one hundredth anniversary of Mercer University Law School in Macon, Georgia.

During his presidential term, President Jimmy Carter visited Robins AFB fourteen times. His visits were so frequent, a special detachment was established at Robins to ensure that security and other arrangements were effectively attended to.

The year 1980 will probably go down in the records as the most eventful for visits in Robins' history. In addition to the visits by President Carter, enroute to his home in Plains, the base was visited by the late President Anwar Sadat of Egypt, in August, and by Israeli Prime Minister Menachem Begin, in September.

General and Mrs. Spalding are at Mr. Vinson's home in Milledgville to escort him to Robins AFB on October 20, 1978 for the ceremony designating the finest suite of the Visiting Officers Quarters as the "Carl Vinson Suite."

Representative Hebert of Louisiana, Chairman of the House Armed Services Committee, is greeted by General Hails in 1973.

U. S. Senator Sam Nunn confers with General Hails in 1973. Senator Nunn, of Perry, Georgia, serves on the Senate Armed Services Committee and is considered authoritative in defense matters.

President Carter transitted Robins on visits to his home in Plains. Air Force One *was hangared at Robins while the Marines provided helicopter transportation between Robins and Plains.*

Maj. Gen. Paulk greets President Carter and Vice President Mondale on a night arrival at Robins. Jody Powell and Hamilton Jordan are in the background.

Robins—Home of Winners

Gen. James R. Allen, Commander-in-chief of the Military Airlift Command and Maj. Gen. John R. Paulk, WR-ALC Commander, inspected the last C-141 to be stretched, "ahead of schedule and under cost," during ceremonies at the Lockheed-Georgia plant in Marietta on June 29, 1982.

Maj. Gen. John R. Paulk, center, was cited in a resolution by the Georgia Senate for support to industry in Georgia. Gov. George Busbee of Georgia signs the resolution as Robert F. Beckmann, WR-ALC small business director and General Paulk look on.

Pictured with the Omaha Trophy, left to right, are: Gen. Bennie L. Davis, SAC Commander-in -chief, Charles W. Durham, Omaha Award Committee, and Col. Richard L. Purdom, 19th Bomb Wing Commander.

Sgt. Rick Trull (left) and SSGt Terry Chambers (right) display trophies won in 1981 by the Robins based 19th Bomb Wing. Left to right, the trophies are: Best Security Police Unit in SAC Award; Best Aircraft Maintenance in SAC; Barrentine Trophy for winning the Giant Sword competition at Ellsworth AFB, South Dakota; Omaha Trophy for the most outstanding SAC Wing; Gen. George C. Kenney Award for Best ORI/Buy None Results; and the Spirit Bell Award, won at the bomb-loading competition at Ellsworth for displaying the best professionalism and greatest esprit de corps.

Lt. Gen. Robert Herres, Eighth Air Force Commander presents the Massey Award to Lt. Col. Richard L. Farnam, chief, 19th Bomb Wing Operations Plans Division.

Colonel Leroy Thornal admires the Outstanding Unit Award that was presented to the 2853ABG for outstanding performance for the period January 1, 1979 to December 31, 1980.

Gen. James P. Mullins, AFLC Commander, presents the Small and Disadvantaged Business Achievement Award to General Paulk.

Colonel Tygett stands in front of the sign displaying the Air Force Organizational Excellence Award for exceptionally meritorious service awarded to the Directorate of Distribution for the period July 1, 1975 to December 31, 1979.

An award was established in 1982 at North Carolina State University in honor of Brig. Gen. W. M. Shaw, Jr., vice commander of WR-ALC. General Shaw was the first graduate of the AFROTC at North Carolina State.

Colonel Richard F. Gillis points to the "Top Directorate of Maintenance" plaque while Earvin L. "Pete" Joyner holds the trophy Maintenance won in 1977 as the top depot maintenance activity in the Air Force. Robins Maintenance personnel have won the AFLC Maintenance award three times since 1970.

Billy Trent, 11 (left), son of TSgt. and Mrs. William Trent, became a hero when he saved the lives of David Lombardi, 7, and his sister, Alicia, 2, after their house caught fire. Billy is pictured here with both the children and their mother, Linda Lombardi.

Children from the Child Care Center got free rides at the Base Exchange Carnival. Here, Lt. Col. Joseph A. Colleran, Jr., Deputy Base Commander, lends a helping hand.

The U. S. Air Force Reserve Band stays busy as it participates in numerous functions throughout the Southeast United States and the Caribbean.

Maj. Larry D. McConnell, chief of base security police, presents a certificate of appreciation to Warner Robins Police Officer Sgt. Jurley R. Barrs for his efforts in crime prevention.

CMSgt. Jack H. Steed (right) retired in June 1982 after 33 years service. Here he receives one of many presentations. CMSgt. Bob Rogers (left) of the Air Force Logistics Command presents him with a shadow box of awards.

The Macon Chamber of Commerce, prime mover in the establishment of Robins Air Force Base, commemorated the 40th Anniversary of Robins by holding its 1981 annual meeting in the east dock of Base Operations, Building 110. Dinner for over one thousand persons was served amidst displays of aircraft and systems managed and flown at Robins Air Force Base.

Robins Federal Credit Union was organized in 1954 with 13 charter members. Its membership has increased to over 56,000 with assets totaling more than $118 million. It has the largest membership of any Credit Union in the State of Georgia. It has four offices including the one above located on Watson Boulevard in the City of Warner Robins, which was completed in 1976.

The 3503d U. S. Air Force Recruiting Group at Robins was voted Number One in the country for 1978, 1979, and 1980.

The 2853d Air Base Group Security Police Squadron at Robins has been named the Outstanding Security Police Unit in AFLC from 1978 to 1981.

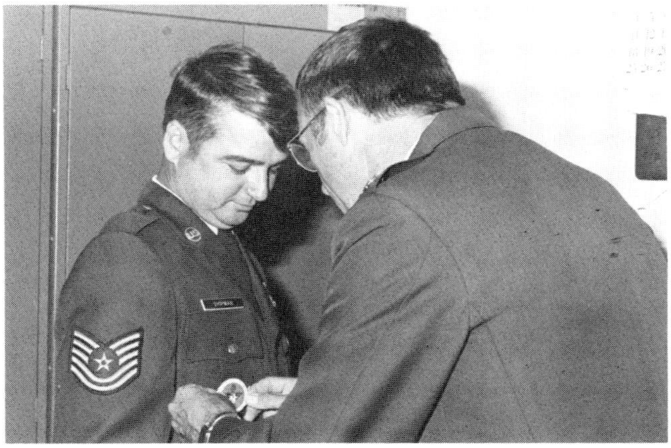

TSgt. James L. Shipman, left, of the 321st Field Training Detachment, receives the Master Instructor Certificate, for an outstanding job of training military personnel, from Col. Leroy W. Thornal, Robins AFB Commander.

Tom Christian, President of the Robins chapter of the Society of Logistics Engineers, presents the SOLE award to Maj. Larry Miller of HQ Air Force Reserves as John Goodrum, SOLE executive director, looks on.

Letty Fitch, naturalist and director of the Spalding Nature Center at Robins, talks with a group of children. Letty has been nominated for the AFLC Natural Resources Conservation Award, the Gen. Thomas D. White Environmental Award.

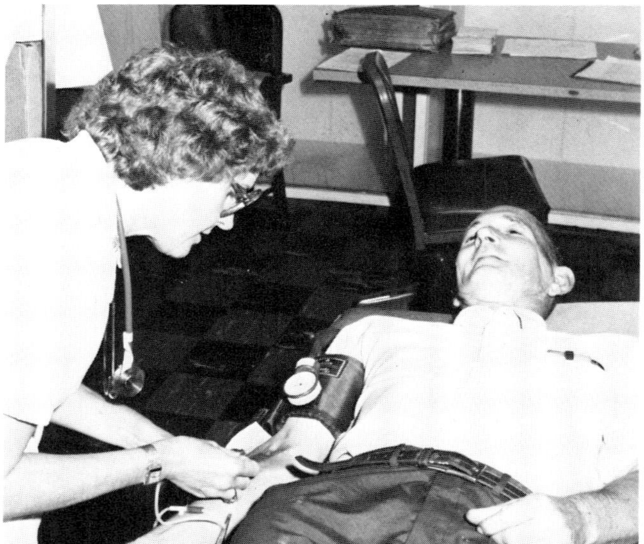

John P. Busby of Maintenance gives his 13th gallon of blood to the Red Cross Bloodmobile at Robins.

Raising the Bicentennial Flag in 1976.

A PHOTOGALLERY: *Leaders of the 1970s*

Commanders, Warner Robins AMA/ALC

Major General Robert E. Hails
August 1972—August 1974

Robert E. Hails was born in Miami, Florida on January 20, 1923. He received a BS degree from Auburn University, Auburn, Alabama, in 1947 and an MS from Columbia University, New York, in 1950.

Hails enlisted in the reserve forces in December 1942 and was commissioned and received his pilot wings in May 1944. During World War II he served with the Thirteenth Air Force in the Pacific Theater. He was on active duty from April 1946 to October 1947 when he was assigned to SAC as an aircraft commander with the 43d Bombardment Wing and then the 2d Bombardment Wing.

In September 1953 he was assigned to HQ United States Air Forces in Europe at the American Embassy in Paris. In 1956 he was assigned to HQ USAF, Washington, D. C., as the Air Staff Program Manager for the F-104 and F-105. General Hails became Commander, San Francisco Air Procurement District, Oakland, California in August 1960. In July 1962 he returned to HQ USAF as Military Assistant for Weapon System Management.

In August 1968, Hails was assigned to HQ AFLC, Wright-Patterson AFB, Ohio, as Assistant Deputy Chief of Staff for Maintenance-Engineering. In August 1971 he assumed command of the Defense Personnel Support Center, Philadelphia, Pennsylvania.

He left Robins to become Vice Commander of Tactical Air Command at Langley AFB, Virginia. He retired as Deputy Chief of Staff for Systems and Logistics at HQ AFLC, Wright-Patterson AFB, Ohio.

He and his wife, Ethel Gayle, have four children: Robert, Merrily, Troy, and Laura. General and Mrs. Hails now live in Dallas, Texas.

Major General Ralph T. Holland
August 1974—September 1975

Ralph T. Holland was born on November 16, 1920 in Cedartown, Georgia. He attended West Georgia College, Carrollton, Georgia prior to enlisting in the U. S. Army Air Corps as an aviation cadet in August 1941. He graduated from advanced flying school in March 1942. In 1947 he returned to college and graduated from Emory University, Atlanta, Georgia.

After graduating from flying school Holland served as an instructor pilot until November 1943, when he joined the 462d Bombardment Group, Hays, Kansas. The group transferred to India and for the next year he flew as a B-29 commander in the China-Burma-India Theater. The 462d Group moved to Tinian Island in the Western Pacific in March 1945.

Holland was on inactive duty from June 1945 to March 1951 when he was recalled to active duty with the 2d Bombardment Wing, Hunter Air Force Base, Georgia, as a deputy squadron commander. He was assigned as Vice Commander of SAC in 1959 and later became Commander of the 341st Bombardment Wing at Dyess AFB, Texas. He became Deputy Commander of SAC's 3d Air Division at Andersen AFB, Guam in June 1965. In July 1968 he was assigned as Commander of the 810th Strategic Aerospace Division, Minot AFB, North Dakota, and in August 1969 became Vice Commander of WRAMA.

General Holland was transferred to the Republic of Vietnam in November 1971 and assumed duties as Deputy Chief of Staff, Logistics, Headquarters Seventh Air Force, at Tan Son Nhut Airfield, and in September 1972 became Deputy Commander of the Seventh Air Force. He became Deputy Chief of Staff for Logistics, HQ PACAF, at Hickam AFB, Hawaii, in February 1973.

General Holland retired from his assignment as WR-ALC Commander. He is presently employed with the Northrop Corporation and he and his wife, Kathleen, live in Hawthorne, California.

Major General William R. Hayes
August 1975—August 1977

William R. Hayes was born in Yonkers, New York on April 5, 1921. He studied business administration at the University of California at Riverside and the University of Omaha, Nebraska. In October 1942 he entered military service as an aviation cadet and in July 1943 received a commission and his pilot wings.

During World War II, Hayes was a B-24 pilot with the 448th and the 93d Bombardment Wings in the European Theater of Operations. He was on inactive duty from November 1945 to March 1951 when he was assigned to SAC as a maintenance squadron commander at March AFB, California. In May 1952 he was assigned to the 98th Bombardment Wing at Yokota Air Base, Japan. In June 1958 he went to SAC headquarters at Offutt AFB, Nebraska as Chief of the Aircraft Branch. He was appointed Vice Commander of the 9th Strategic Reconnaissance Wing at Beale AFB in October 1966 and in December 1966 assumed command.

In May 1972 he became Deputy Director for Logistics for HQ U. S. European Command at Stuttgart, Germany. General Hayes was assigned as Director of Maintenance, Engineering, and Supply in the office of the Deputy Chief of Staff, Systems and Logistics, HQ USAF, in May 1974.

General Hayes retired from the position of WR-ALC Commander in August 1977. He is married to the former Ann Hannan and they have four sons: Thomas, Timothy, Robert and Daniel; and a daughter Susan. General and Mrs. Hayes live in Fair Oaks, California.

Major General John R. Spalding, Jr.
August 1977—January 1980

General Spalding was born in Oak Park, Illinois, and graduated from high school in Downers Grove, Illinois. He received a Master's degree in 1958 from the University of Chicago School of Business under the Air Force Institute of Technology program.

Spalding was commissioned and rated a navigator in November 1944. In 1946 he was assigned to the Alaskan Air Command. In 1949 he graduated from the single-engine fighter school at Williams AFB, Arizona. The next two years he served as a fighter pilot with the 93d Fighter Squadron at Kirtland AFB, New Mexico. In 1951 he transferred to the 51st Fighter Group at Suwon, Korea. In 1952, Spaulding was assigned to Luke AFB, Arizona, as a fighter gunnery instructor. In February 1954, he was selected as the left wingman of the Thunderbirds.

In 1959 he was assigned to SAC and served at Dyess AFB, Texas, where he progressed to Assistant Deputy Commander for Operations of the 96th Strategic Aerospace Wing. In June 1968 General Spalding became Deputy Commander for Operations of the 405th Fighter Wing at Clark AB, Philippines. In 1970 he went to Vietnam as Vice Commander and later Commander of the 366th Tactical Fighter Wing "The Gunfighters." The wing was awarded the General Robert Worley Trophy as the best fighter wing in Southeast Asia in 1971.

In 1975 he became Deputy Chief of Staff, Logistics for North American Air Defense Command/Continental Air Defense Command/Aerospace Defense Command, Ent AFB, Colorado. He held this position until he was assigned as WR-ALC Commander.

General Spalding retired from his position as WR-ALC Commander. General Spalding has four children: Larry, Robert, Bruce, and Trena. He and Mrs. Spalding (the former Elizabeth T. Light) live in Colorado Springs, Colorado.

Vice Commanders, HQ AFRES at Robins AFB

Brig. Gen. Alfred Verhulst
January 1972—May 1973

General Verhulst was born in Sheboygan Falls, Wisconsin on March 7, 1921. He attended Sheboygan Business College. He served with the Air Corps at Aloe Field, Texas, Coffeyville, Kansas, and New Guinea. He received the Air Medal after the Luzon Campaign, Corregidor and Noemfoor invasions. His reserve affiliation began at Milwaukee, Wisconsin, in June 1952. In 1956 he was appointed Commander of the 438th Fighter Bomber Group. In 1958 he was assigned to the 440th Troop Carrier Wing and in 1959 served on the Air Reserve Airlift Force operations staff. In 1961, his aircrew won two of the three events in the CONAC National Troop Carrier Competition. This was the first time in the history of the competition that one aircrew won both events. In 1961 he served as 440th Troop Carrier Wing Swift Strike Mission Commander. He was the 1206th Troop Carrier Wing (Provisional) Commander for the 1963 Swift Strike Exercise.

General Verhulst died in 1975.

Maj. Gen. Earl O. Anderson
May 1973—October 1976

General Anderson was born October 5, 1924 in Mead, Nebraska. He attended York Junior College, the University of Arizona and received a bachelor of science degree from the University of Omaha. He was commissioned in January 1944. His first assignment was at Liberal Army Airfield, Kansas. During World War II he served at New Guinea with the Thirteenth Air Force. From 1947 to 1948 he was on inactive duty. He served with SAC in key staff and command positions, including HQ SAC at Offutt AFB, Nebraska and the 7th Air Division HQ in England, from 1948 to 1955. In 1958 he was assigned as deputy wing commander of the 512th Troop Carrier Wing at Willow Grove, Pennsylvania. Following his service at Robins, Anderson retired and lives in Marietta, Georgia.

Maj. Gen. Richard Bodycombe
November 1976—April 1979

General Bodycombe was born April 29, 1922, in Pittsburgh, Pennsylvania. He received bachelor and master of science degrees from the University of Michigan. He was commissioned in May 1944 and served on active duty at Turner Field, Georgia, Davis-Monthan AFB, Arizona, and in Europe. He returned to the United States in 1952 and was assigned to Selfridge AFB, Michigan, Altus AFB, Oklahoma, and Donaldson AFB, South Carolina. In 1955 he was selected for duty at the U. S. Air Force Academy.

General Bodycombe reverted to Reserve status in February 1956 and served at Chicago-O'Hare International Airport, Selfridge AFB, Michigan, Ellington AFB, Texas, Barksdale AFB, Louisiana, and Offutt AFB, Nebraska. He was recalled to active duty in November 1976 to become vice commander of HQ AFRES at Robins. General Bodycombe is currently chief of Air Force Reserve, at HQ USAF.

General Bodycombe is married to the former Karin Lindh and they have three sons, Brian, Peter and Chris.

Vice Commanders, Warner Robins AMA/ALC

Brig. General John C. Bartholf
November 1971—June 1973

Brig. General Garry A. Willard, Jr.
July 1973—July 1974

Brig. General Carl G. Schneider
August 1974—March 1976

Brig. General William J. Becker
April 1976—August 1978

Brig. General Marvin C. Patton
August 1978—June 1980

Biographical Notes on the WRAMA/ALC Vice Commanders:

Brigadier General Bartholf: General Bartholf was born in the Panama Canal Zone on September 30, 1925. Graduating from the United States Military Academy at West Point, NY, in 1946, he served at Williams AFB, AZ; at Shaw Field, SC, at Keesler AFB, MS, at HQ USAF, as detachment commander of a special project in Brazil, at McClellan AFB, CA, and at Nellis AFB, NV. General Bartholf participated in the Armed Forces Pilot Exchange Program with the Navy and was authorized to wear US Navy pilot wings. He also served at the U.S. Air Force Academy and at Lowry AFB, CO; at Ethan Allen AFB, VT; with HQ ADC, at Ent AFB, CO.; in Europe at Ramstein and Woodbridge; at Davis-Monthan AFB, AZ; and then at Ubon and Udorn Airfields in Thailand. He served at HQ USAF prior to a return to Europe at Bentwaters and Wiesbaden. After his tour at WRAMA, he transferred back to HQ USAF. General Bartholf is married to the former Patsy Adams of Sacramento, CA. They have three children: Paige, John and Mark. He is retired and resides in Alexandria, VA.

Brigadier General Willard: General Willard was born in Utica, NY on May 10, 1928. He entered the Army in 1944 and served with the Korean Occupation Forces in Korea in 1946. Returning to active duty as an aviation cadet, he served as a fighter pilot during the Korean War; served as a flight commander and operations officer; served two combat tours in Southeast Asia; served at HQ USAF in the "Wild Weasel" program and as a project officer of the new F-15 air superiority fighter. Before coming to Robins, General Willard served as Vice Commander of the 314th Air Division in Korea and as Vice Commander and Commander of the F-105 equipped "Flying Tigers" of the 23rd Tactical Fighter Wing. General Willard is retired and lives overseas.

Brigadier General Schneider: General Schneider was born in Ralls, TX, on March 6, 1928. He entered the U.S. Army Air Forces, then entered aviation cadet training. He served in the Far East during the Korean War, at Luke AFB, AZ, at Nellis AFB, NV, at Maxwell AFB, AL, at Laughlin AFB, TX, returned to Nellis, was assigned to Bitburg Air Base, Germany, returned to Luke, and served in Vietnam. He served a third time at Luke, at MacDill AFB, FL, and then at HQ USAF, prior to his first Georgia tour at Moody AFB. He served in Korea prior to his tour at Robins. Major General Schneider was named Commander, Oklahoma City ALC, and served in the Command Section at HQ AFLC. He is married to the former Elaine Eliason, they have a son, Robert, and a daughter, Debra. He is retired and living in Scottsdale, AZ.

Brigadier General Becker: General Becker was born on March 14, 1926, in New York City, NY. After serving in World War II in Europe, he received his commission through the ROTC program at the University of Southern California. He served in Korea, at Shaw AFB, SC, in Japan, returned to Shaw, at Langley AFB, VA; at Ramstein AB, Germany; in Southeast Asia; and at HQ USAF. Following a tour of duty at Nellis AFB, NV, General Becker served at HQ AFSC and then at Robins. General Becker is married to the former Esther Krueger of Denver, CO. They have three children: Mark, William and Lisa. He transferred to Battle Creek, MI, with DLA. He is now retired and lives in Las Vegas, NV.

Brigadier General Patton: General Patton was born on December 4, 1929, in Colorado Springs, CO. He graduated from the Colorado State University, was commissioned and then trained at Bainbridge AFB, GA, Laredo AFB, TX, and Nellis AFB, NV. He was assigned in Korea, with ADC at Geiger Field, Spokane, WA, at the Air Force Academy, at Webb AFB, TX, at Luke AFB, AZ, in Vietnam, at HQ USAF, at RAF Lakenheath, England and Ramstein AB, Germany, and at Oklahoma City ALC, Tinker AFB, OK. After his tour at Robins, he was reassigned as Executive Director of Supply Operations, DLA, Cameron Station, and was then named HQ USAF Comptroller. He retired in the summer of 1982. General Patton is married to the former Ardyce J Froid of Denver, CO. He has one son, Scott, and one daughter, Robin.

WRAMA/ALC Chiefs of Staff/ Executive Assistants/Directors/Deputies and Tenant Commanders

Col. Howard G. Beck
Chief of Staff
September 1970—May 1972

Col. James J. Rouch
Chief of Staff
September 1972—January 1973

Col. Albert L. Neuhauser
Chief of Staff
January 1973—October 1973

Col. Joseph E. Cahill
Chief of Staff
November 1973—August 1975

Col. Sidney E. Moody
Chief of Staff
August 1975—January 1976

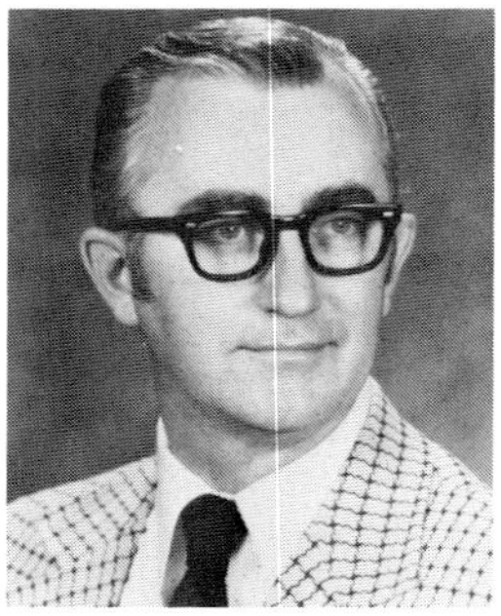

Mr. Robert L. Burns
Executive Assistant
October 1976—August 1981

Col. Harry Z. Moore
Director of Distribution
May 1970—June 1971

Col. Ray R. Hewitt
Director of Distribution
October 1971—November 1972

Col. James K. Lowman
Director of Distribution
December 1972—June 1975

Col. James W. Dennis, Jr.
Director of Distribution
June 1975—March 1979

Col. Verne B. Hildebrant
Director of Distribution
April 1979—September 1980

Mr. Gerald L. Tompkins
Deputy Director of Distribution
August 1971—March 1974 and
April 1975—January 1981

Mr. John W. Clark
Deputy Director of Distribution
March 1974—April 1975

Col. Roka Pamperien
Director of Maintenance
April 1972—February 1973

Col. James E. Burkhalter
Director of Maintenance
February 1973—June 1975

Col. James R. Landry
Director of Maintenance
June 1975—February 1978

Col. Robert D. Smith
Director of Maintenance
February 1978—August 1980

Mr. George Calver
Deputy Director of Maintenance
April 1970—August 1975

Mr. Earvin L. Joyner
Deputy Director of Maintenance
September 1975—Present

Col. Irving B. Schoenberg
Director, Materiel Management
September 1970—June 1973

Col. Jay T. Edwards, III
Director, Materiel Management
July 1973—July 1975

Col. Leo Marquez
Director, Materiel Management
August 1975—August 1977

Col. Donald P. Litke
Director, Materiel Management
August 1977—February 1979

Col. Charles A. Kelly
Director, Materiel Management
March 1979—June 1981

Col. Jacob B. Pompan
Director of Procurement and Production
June 1970—June 1973

Col. Thomas F. Blake, Jr.
Director of Procurement and Production
June 1973—June 1976

Col. James C. Dever, Jr.
Director of Procurement and Production
July 1976—May 1979

Col. John E. Elliott
Director
Contracting and Manufacturing
June 1979—Present

Mr. James M. Mischo
Deputy Director
Procurement and Production
April 1973—July 1977

Mr. Lester M. Carter
Deputy Director
Contracting and Manufacturing
August 1977—Present

Col. Berry A. Seagraves, Jr.
Director of Plans and Programs
June 1970—May 1971

Col. J. J. Rouch
Director of Plans and Programs
June 1972—October 1972

Col. James W. Bailey
Director of Plans and Programs
October 1972—July 1973

Col. Sidney E. Moody
Director of Plans and Programs
August 1973—August 1975

Col. Roy L. Gentry, Jr.
Director of Plans and Programs
August 1975—August 1976

Col. Robert L. Bobbett
Director of Plans and Programs
August 1976—July 1978

Col. Terry P. Weyant
Director of Plans and Programs
July 1978—August 1980

Mr. Lucian M. Rucker
Deputy Director of Plans and Programs
August 1976—August 1981

Col. Frederick F. Shriner
Base Commander
July 1973—August 1974

Col. Stillman V. Taylor
Base Commander
September 1974—May 1975

Col. Edward Young
Base Commander
May 1975—October 1976

Col. Leslie C. Candee
Base Commander
October 1976—June 1978

Col. Fred E. Pirkey
Base Commander
August 1978—August 1979

Col. Howard W. McWhorter
Base Commander
August 1979—April 1980

Col. Robert E. Stevens
Director of Personnel
July 1970—April 1973

Col. Eugene M. Bushong
Director of Personnel
December 1974—June 1979

Col. Eleanor L. Skinner
Director of Personnel
June 1979—Present

Mr. Ray W. Kelso
Civilian Personnel Officer and
Deputy Director of Personnel
July 1970—July 1972

Mr. Richard A. Morrissey
Civilian Personnel Officer and
Deputy Director of Personnel
February 1975—June 1982

Col. Edward D. Young
Comptroller
August 1973—April 1975

Col. Kenneth A. Simonet
Comptroller
June 1975—June 1978

Col. Ralph Law
Comptroller
September 1978—November 1981

Mrs. Vivian P. Hartman
Deputy Comptroller
1979—Present

Col. Nathaniel A. Gallagher
19th Bomb Wing Commander
March 1973—February 1974

Col. Frank H. McArdle
19th Bomb Wing Commander
February 1974—July 1975

Col. Edwin L. Smith
19th Bomb Wing Commander
July 1975—June 1977

Col. Charles W. Reeves, Jr.
19th Bomb Wing Commander
June 1977—June 1978

Col. Donald O. Pflugrath
19th Bomb Wing
Commander
June 1978—January 1980

Col. George L. Madara
5th Mobile Communications
Group Commander
July 1971—August 1973

Col. Clifford Allison, Jr.
5th Mobile Communications
Group Commander
September 1973—August 1974

Col. Patrick J. Frickleton
5th Mobile Communications
Group Commander
September 1974—September 1975

Col. Robert S. Rayfield
5th Mobile Communications
Group Commander
September 1975—March 1978

Col. John Kopsick, Jr.
5th Mobile Communications
Group Commander
March 1978—May 1981

Biographical Notes on the WRAMA/ALC Chiefs of Staff/ Executive Assistants/Directors/Deputies and Tenant Commanders:

Col. Beck (See previous chapter.)

Col. James J. Rouch: Colonel Rouch is a native of Fort Sheridan, IL. He had served as Director of Plans and Programs before being named Chief of Staff. He retired in January 1973.

Col. Albert L. Neuhauser: Colonel Neuhauser is a native of NY. His previous assignments at Robins were in Maintenance and as Inspector General. He retired on October 31, 1973.

Col. Joseph E. Cahill: Colonel Cahill was born in Leavenworth, KS. His previous assignments at Robins were as Deputy Base Commander and Director of Plans and Programs. Colonel Cahill retired in August 1975.

Col. Sidney E. Moody: Colonel Moody was a native of Canton, GA. He came to Robins in July 1973 as Director of Plans and Programs. He left Robins to become the Assistant for Logistics Plans at HQ ATC. He is now retired and lives in San Antonio, TX.

Mr. Robert L. Burns: Mr. Burns came to Robins from HQ AFLC. He is currently serving as Director of Plans and Programs at Robins.

Col. Harry Z. Moore: Col Moore assumed Command of the Defense Property Disposal function in Memphis, TN, upon his departure from Robins. He retired from the Air Force in 1974 and currently lives in Memphis, TN.

Col. Ray R. Hewitt: Colonel Hewitt instituted a daily Stand-Up Briefing to be more fully aware of the status of operations in the Directorate; he established an NCOIC to handle military affairs; and improved methods of storage by eliminating much of the outside storage locations. Retiring from this position at Robins, he began a farming operation in Rochville, NE.

Col. James K. Lowman: Col Lowman implemented a successful motivational program emphasizing "People are important," called locally the MIDI (Mission Is Distribution Identification) Program. He led in the completion of the Logistical Materials Processing Facility, Building 376. Col. Lowman is currently Assistant Deputy Chief of Staff for Logistics Operations at Headquarters Air Force Logistics Command.

Col. James W. Dennis, Jr.: Col. Dennis led in the completion of the Avionics Warehouse, which features the most advanced warehouse system on base: the Warehouse Information Control System, which receives and stores information and interfaces with other computer management systems. During his tenure, the Directorate received the highest rating ever received from the AFLC Inspector General during their annual management inspection. Colonel Dennis currently serves as Commander of the Defense Property Disposal Region—Europe, a position he assumed following transfer from Robins.

Col. Verne B. Hildebrant: Col. Hildebrant officially opened the Consolidation/Containerization Point function, a totally new concept for the Air Force. All directed low priority surface shipments throughout the United

States destined for selected bases across the Atlantic are received here and consolidated into van type containers; these are then moved by commercial carrier lines to seaports in Savannah or Charleston. During his tenure the Directorate received a second rating of "Excellent" from the AFLC Inspector General. Following his transfer from Robins, Colonel Hildebrant served as Director of the Storage and Transportation Division, and later became Vice Commander of the Defense Construction Supply Center at Columbus, OH. He now lives in Pickerington, OH.

Mr. Gerald L. Tompkins: Mr. Tompkins implemented the first Equal Employment Opportunity and Federal Women's Programs in the Directorate. He led in the formulation of the Warehouse Information Control System in the Avionics Warehouse. During his tenure, the AFLC Inspection Team ranked the directorate as "Excellent," with seven specific accomplishments highlighted by the team, the highest rating ever achieved. Mr. Tompkins' break in service at Warner Robins occurred during a transfer to Vietnam; he transferred to Sacramento Air Logistics Center in January 1981 as Deputy Director of Distribution.

Mr. John W. Clark: Following a period of service as Chief of the Management Services Division, Mr. Clark served as Deputy Director during the transfer of Mr. Tompkins to Vietnam. He led in the successful start-up of the massive new Logistical Material Processing Facility and the Avionics Warehouse. Mr. Clark transferred in June 1981 to the position of Chief of the Depot Operations Division at Headquarters, Air Force Logistics Command.

Col. Roka Pamperien: During his tenure, the first direct support of the F-15 aircraft occurred, and Robins was assigned Technology Repair Center responsibility for C-130, C-141, and F-15 aircraft, airborne electronics, gyros, life support equipment, propellers, and portable logistics buildings. Colonel Pamperien is retired and resides in Warner Robins, GA.

Col. James E. Burkhalter: During this period, maintenance was performed on C-141 Starlifter and seventeen different configurations of the C-130 Hercules. Under his leadership, Maintenance again won top honors in 1973 as the best work center in the command. Airborne Electronics Division was recognized as the "Avionics Center of the Air Force," the largest product division in AFLC. Colonel Burkhalter is retired and lives in Roswell, GA.

Col. James R. Landry: The F-15 Speedline was set up during this period to retrofit power systems; all outstanding time compliance technical orders were effected while aircraft power systems were being modified. To conserve energy, a solar purge fluid reclamation distillery became operational in 1977. Under Colonel Landry's guidance, the directorate won the AFLC Maintenance Award for top performance in the command. Colonel Landry left Robins to become the Director of Logistics HQ TUSLOG Ankara, Turkey. In August 1980, he became Assistant to the Commander for Quality Assurance at HQ AFLC, Wright-Patterson AFB, OH. He retired on January 31, 1982 and is now living in Covington, GA.

Col. Robert D. Smith: A significant workload during this time involved the receipt of the first C-141B Starlifter from Military Airlift Command. A new laboratory for repair of *Pave Spike* laser weapons was opened in the Avionics Center in June 1980. Colonel Smith left Robins for an assignment in Ramstein, Germany. He is currently assigned to HQ AFLC as Director of Propulsion Systems in the Logistics Operations area.

Mr. George Calver: Under his guidance, more than 280 aircraft were processed through repair shops during nine months in 1970. Repair flow time was cut and aircraft were produced "better, cheaper, faster." The directorate won the AFLC Maintenance Award for top performance in the command in 1971 and again in 1973. Mr. Calver is retired and resides in FL.

Mr. Earvin L. Joyner: Mr. Joyner has led in achieving the highest levels of productivity among Maintenance Directorates of the Command. He has emphasized people programs with great effectiveness, a strong reason for his successes in increasing productivity.

Col. Irving B. Schoenberg: Colonel Schoenberg retired from the Air Force from his position as Director of Materiel Management at WR-ALC. He presently is President of Abrams Industries, makers of display fixtures for the merchandising industry and lives in Atlanta, GA.

Col. Jay T. Edwards, III: Following Colonel Edwards' service at Robins, he served as Commander of the Oklahoma City Air Logistics Center, a postion from which Major General Edwards retired on July 31, 1982.

Col. Leo Marquez: Following his service at Robins, Marquez transferred to the position of Deputy Chief of Staff for Plans and Programs at Headquarters AFLC. Major General Marquez currently (July 1982) serves as Commander of the Ogden Air Logistics Center.

Col. Donald P. Litke: Following his service at Robins, Litke transferred to the position of Vice Commander, Oklahoma City Air Logistics Center. Currently (July 1982), Brigadier General Litke serves as Deputy Director, Logistics and Security Assistance, Headquarters European Command and is stationed at Vaihingen, Germany.

Col. Charles A. Kelly: Colonel Kelly retired from his position as Director of Materiel Management at WR-ALC. He is currently working with Minneapolis-Honeywell in Tampa, FL.

Col. Jacob B. Pompan: Colonel Pompan is retired. He has a large law practice in Washington, D. C.

Col. Thomas F. Blake, Jr.: During Colonel Blake's tour of duty the directorate awarded a multi-million dollar contract for modification and testing of a prototype C-141 aircraft as part of the C-141 stretch program, with contract options to modify the remainder of the C-141 fleet. He was also instrumental in the development of a new management system for measuring personnel effectiveness. This system has been adopted AFLC-wide. Colonel Blake left Robins to become Chairman of the Armed Services Procurement Regulation Committee (ASPR), Office of Assistant Secretary of Defense (I&L), Washington, D. C. He has now retired from the Air Force and is engaged in the construction business in Washington, D. C., living in Alexandria, VA.

Col. James C. Dever, Jr.: Colonel Dever was promoted to rank of Brigadier General and reassigned as Commander, AFCMC, Kirtland AFB, NM.

Col. John E. Elliott: Colonel Elliott initiated action to expand consolidation of requirements by modifying the J023 Automated Purchase System. Previously, the preparation of an automated "low density" purchase request was limited to a single line item. Based on the modification, multiple line item "low density" purchase requests are mechanically generated. This program was implemented command wide in 1981. It has greatly reduced workload within the contracting directorate and resulted in significant savings for the government.

Mr. James M. Mischo: Two major programs during his tenure were Project Rivet Bat which involved a critical radar warning system for first line fighter and bomber aircraft and represented a program in excess of $27 million, and the award of a contract for the 1050-II computer. The computer program touched 146 bases that provided the critical base supply network for the Air Force. Mr. Mischo transferred to Sacramento Air Logistics Center in July 1977 and is presently serving as Deputy Director, Contracting & Manufacturing at that ALC.

Mr. Lester Carter: In 1967, Mr. Carter represented AFLC and all ALC procurement centers in Washington, D. C. in developing the present contracting and manufacturing organizations and obtaining the present authorized grade structure. He was instrumental in the directorate receiving the USAF Contracting and Acquisition Award in 1979 and initiated the modification of the JO23 Automated Purchase System to expand consolidation of requirements.

Col. Berry A. Seagraves, Jr.: Colonel Seagraves left Robins for an assignment in Southeast Asia. He is now retired and is in the construction business in the Middle Georgia area.

Col. W. F. Seith, Director of Plans and Programs, May 1971—May 1972 (not pictured): Colonel Seith retired in May 1972.

Col. J. J. Rouch: Colonel Rouch served as WRAMA Chief of Staff from October 1972 until his retirement in January 1973. He is currently living in Fort Valley, GA and is an instructor at Fort Valley State College.

Col. James W. Bailey: Colonel Bailey retired on July 31, 1973 and presently lives in FL.

Col. Sidney E. Moody: Colonel Moody was reassigned as the ALC Chief of Staff. He left Robins for an assignment as the Assistant for Logistics Plans at the Air Training Command, Randolph AFB, TX. Colonel Moody is now retired and lives in San Antonio, TX.

Col. Roy L. Gentry, Jr.: Following his service at Robins, he was reassigned to the Air Force Inspector General's Office at Norton AFB, CA. He is now retired and lives in Little Rock, AR.

Col. Robert L. Bobbett: Colonel Bobbett retired on July 20, 1978. He is now living in Las Vegas and is in the real estate business.

Col. Terry P. Weyant: Colonel Weyant was reassigned as Deputy Chief of Staff for Plans, HQ PACAF, Hickam AFB, HI

Mr. Lucian M. Rucker: Mr. Rucker transferred to Robins upon the phase-out of Mobile AMA. Following his service as Deputy Director, he was reassigned as Executive Assistant to the ALC Commander, a position he now holds.

Col. Stillman V. Taylor: Colonel Taylor was reassigned to Hq AFRES.

Col. Leslie C. Candee: Shortly after his appointment, Robins became a reception base for President Carter on his trips between Washington and Georgia. Colonel Candee has retired and is now raising pistachio nuts in Tucson, AZ.

Col. Fred E. Pirkey: Support for President Carter continued through Colonel Pirkey's tenure. Colonel Pirkey retired from the Air Force following a tour of duty at Chanute Air Force Base, IL, and is engaged in real estate in the Atlanta area.

Col. Howard W. McWhorter: Colonel McWhorter also provided support to President Carter. Colonel McWhorter retired from the Air Force at Robins Air Force Base and now resides in Athens, GA.

Col. Robert E. Stevens: Colonel Stevens was reassigned as Personnel Chief at Kelly AFB, TX.

Col. Willard E. Wentz, Director of Personnel, April 1973—December 1974 (not pictured): Colonel Wentz was reassigned as Deputy Chief of Staff, Personnel, HQ AFRES.

Col. Eugene M. Bushong: Colonel Bushong was reassigned to Fort Sheridan, IL as Director of Testing, Air Force Element, Military Enlistment Processing.

Mr. Ray W. Kelso: Mr. Kelso left Robins to become the Director, Civilian Personnel, HQ AFSC, Andrews AFB, Washington, D. C. Following this tour of duty, he accepted a position with the United States Air Force at Loring AFB, MA.

Mr. Jack A. Wecker, Deputy Director of Personnel, September 1972—February 1975 (not pictured): Mr. Wecker was reassigned to Hill AFB, UT in 1975.

Mr. Richard A. Morrissey: Mr. Morrissey retired from civil service at Robins in June 1982.

Col. Edward D. Young: Colonel Young was reassigned as Robins AFB Commander in April 1975 and left Robins in November 1976 to become director of the Senior Officer Education Leadership Management and Development Center, at the Air University, Maxwell, AFB, AL.

Col. Kenneth A. Simonet: Colonel Simonet was reassigned in June 1978 as the Air Force Liaison Officer in Moehrengen, Germany.

Col. Ralph Law: Colonel Law retired in November 1981 to become plant manager of Gould, Inc., an electronics firm in Warner Robins, GA.

Mrs. Vivian P. Hartman: Mrs. Hartman is the highest ranking civilian female employee at Robins. She grew up in the computer science field and has seen the Data Automation activity grow from a punched card accounting machine (PCAM) to one of the largest computer centers in the Southeast. Most of her career has been in the comptroller organization with the exception of ten years spent in the Plans and Management Office.

Col. William R. Thorstenson, 19th Bomb Wing Commander, February 1970—September 1971 (not pictured): Colonel Thorstenson retired in March 1972.

Col. Dudley G. Kavanaugh, 19th Bomb Wing Commander, September 1971—August 1972 (not pictured): Colonel Kavanaugh retired in August 1975.

Col. Stanley C. Beck, 19th Bomb Wing Commander, August 1972—March 1973 (not pictured): Throughout his term Colonel Beck was TDY at Guam, for the Southeast Asia conflict. Also, during his tenure the wing received the SAC Hall of Fame Certificate for twelve years of accident free flying. After leaving the wing, Colonel Beck went to the 2nd Bomb Wing at Barksdale AFB, LA. At present, Maj. Gen. Beck is stationed at Andersen AFB, Guam.

Col. Nathaniel A. Gallagher: Colonel Gallagher retired in June 1977.

Col. Edwin L. Smith: During Colonel's Smith tenure the 19th Organizational Maintenance Squadron received the Air Force Outstanding Unit Award. Colonel Smith retired at Omaha, NE in 1979 after serving as a team chief for SAC/IG.

Col. Charles W. Reeves, Jr.: Colonel Reeves left the wing to become a member of the AFLC Inspector General team at Wright-Patterson AFB, OH.

Col. Donald O. Pflugrath: During Colonel Pflugrath's command the 912th Air Refueling Squadron won the Spaatz Refueling Trophy during the 1978 SAC Giant Voice Bombing and Navigation Competition and the SAC Golden Boom Trophy for the period of July 1 to December 31, 1978. The 912th Air Refueling Squadron, the 19th Field Maintenance Squadron, the 19th Munitions Maintenance Squadron, the 19th Organizational Maintenance Squadron, and the 19th Security Policy Squadron all received Outstanding Unit Awards during Colonel Pflugrath's tenure. Colonel Pflugrath left the wing to go to HQ SAC, Offutt AFB, NE.

Col. George L. Madara: During his tenure, a fixed Digital Subscriber Terminal was installed, tested, and made operational in the headquarters building. The TSW-7, mobile control tower, was also acquired. Colonel Madara is retired and lives in Warner Robins, GA.

Col. Clifford Allison, Jr.: The group received the Air Force Outstanding Unit Award and the Major General Harold M. McClellan Award during Colonel Allison's tenure. The group also provided support to the local community during an unusual and unexpected snowfall. Four wheel drive vehicles were used to transport local hospital employees to work during shift changes. Colonel Allison is retired and lives in VA.

Col. Patrick J. Frickleton: During this period the group assisted local and state authorities in the search for two missing persons. Forty-two personnel and numerous pieces of equipment were used in the operation. Colonel Frickleton is retired.

Col. Robert S. Rayfield: During this period the group received the Air Force Outstanding Unit Award and the Air Force Communications Service Commander's Achievement Award. The group also provided communications equipment and personnel in support of Project *New Horizons*, the Vietnamese Refugee Program, at Eglin AFB, FL. Colonel Rayfield is retired and lives in TX.

Col. John Kopsick, Jr.: During this period the group participated in operation *Proud Phantom*. The group was among the first American military forces to set foot on Eqyptian soil since World War II. Within twenty-four hours after landing in Eqypt, voice and data satellite communications were established with the United States and Europe. In addition, the first application of long distance remoting of a supply computer was successfully demonstrated during the operation. Colonel Kopsick is retired and lives in Warner Robins, GA.

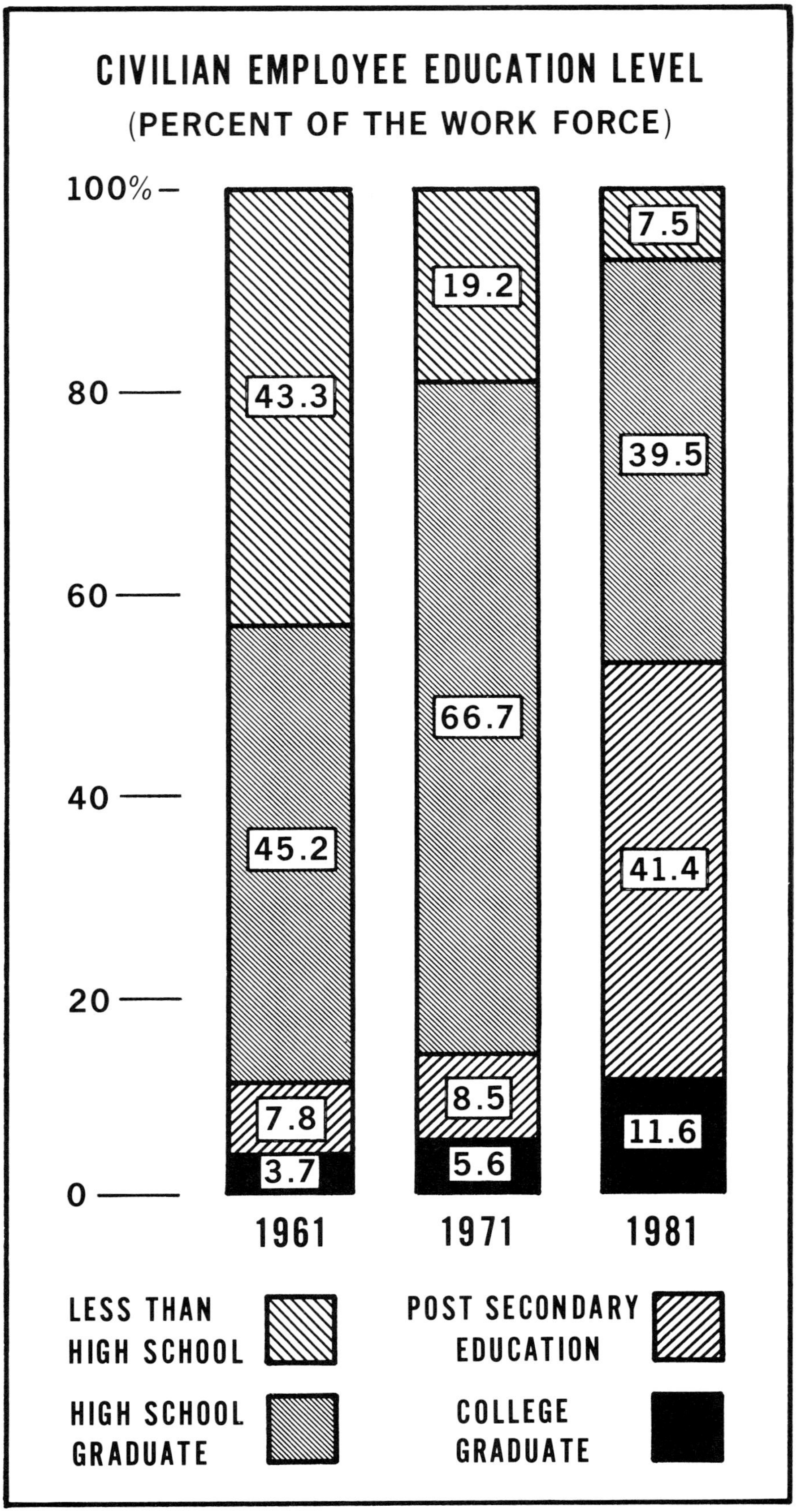
CIVILIAN EMPLOYEE EDUCATION LEVEL
(PERCENT OF THE WORK FORCE)
100%
80
60
40
20
0
43.3
45.2
7.8
3.7
19.2
66.7
8.5
5.6
7.5
39.5
41.4
11.6
1961
1971
1981
LESS THAN HIGH SCHOOL
HIGH SCHOOL GRADUATE
POST SECONDARY EDUCATION
COLLEGE GRADUATE

ROBINS POPULATION

1950 – 1982

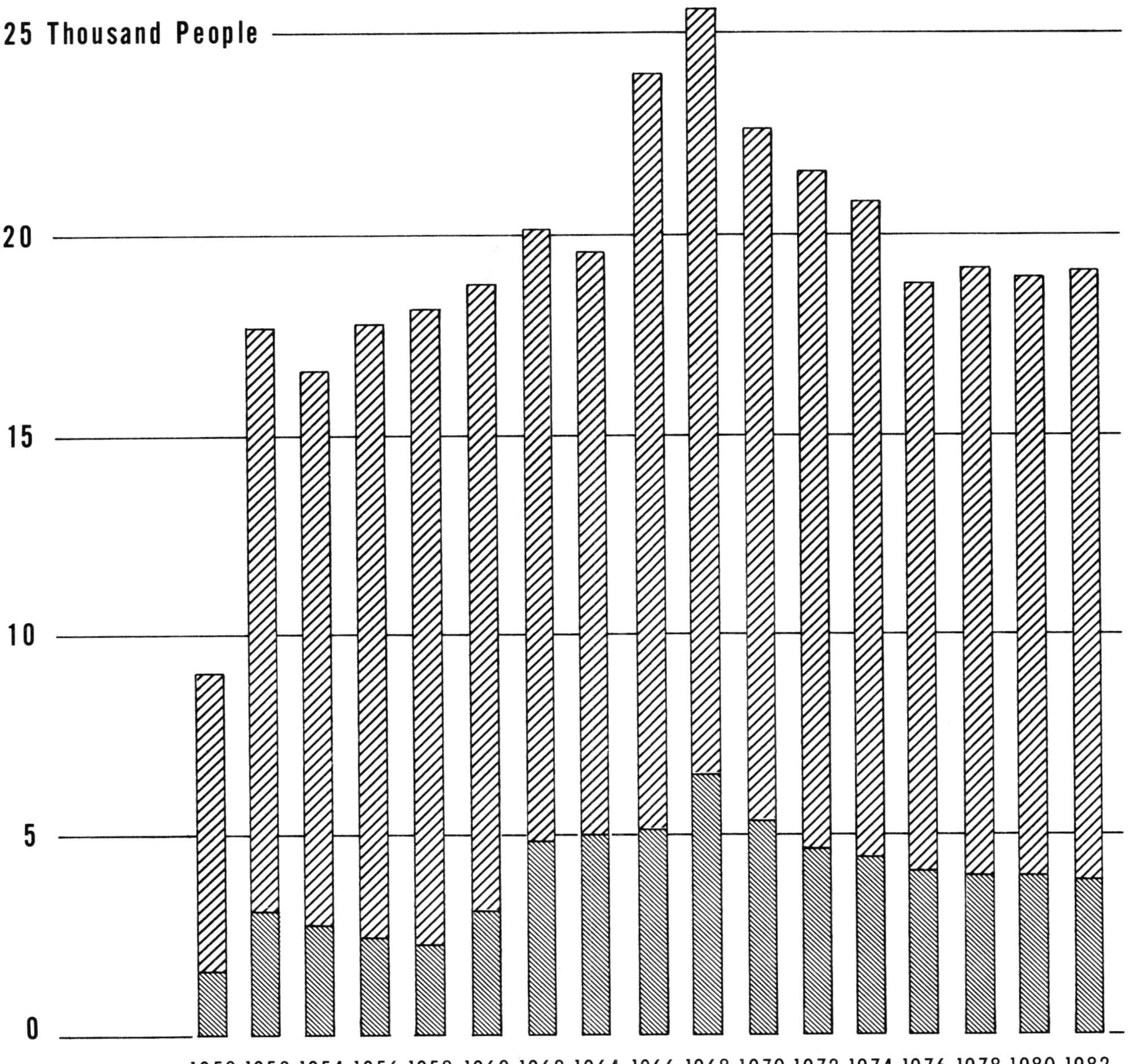

	1950	1952	1954	1956	1958	1960	1962	1964	1966	1968	1970	1972	1974	1976	1978	1980	1982
CIVILIAN	7.4	14.6	13.8	15.3	15.8	15.7	15.2	14.6	18.9	19.1	17.3	16.9	16.4	14.7	15.2	15.0	15.2
MILITARY	1.7	3.1	2.8	2.5	2.4	3.1	4.9	5.0	5.1	6.5	5.4	4.7	4.5	4.1	4.0	4.0	3.9
TOTAL	9.1	17.7	16.6	17.8	18.2	18.8	20.1	19.6	24.0	25.6	22.7	21.6	20.9	18.8	19.2	19.0	19.1

ROBINS PAYROLL

1950 – 1982

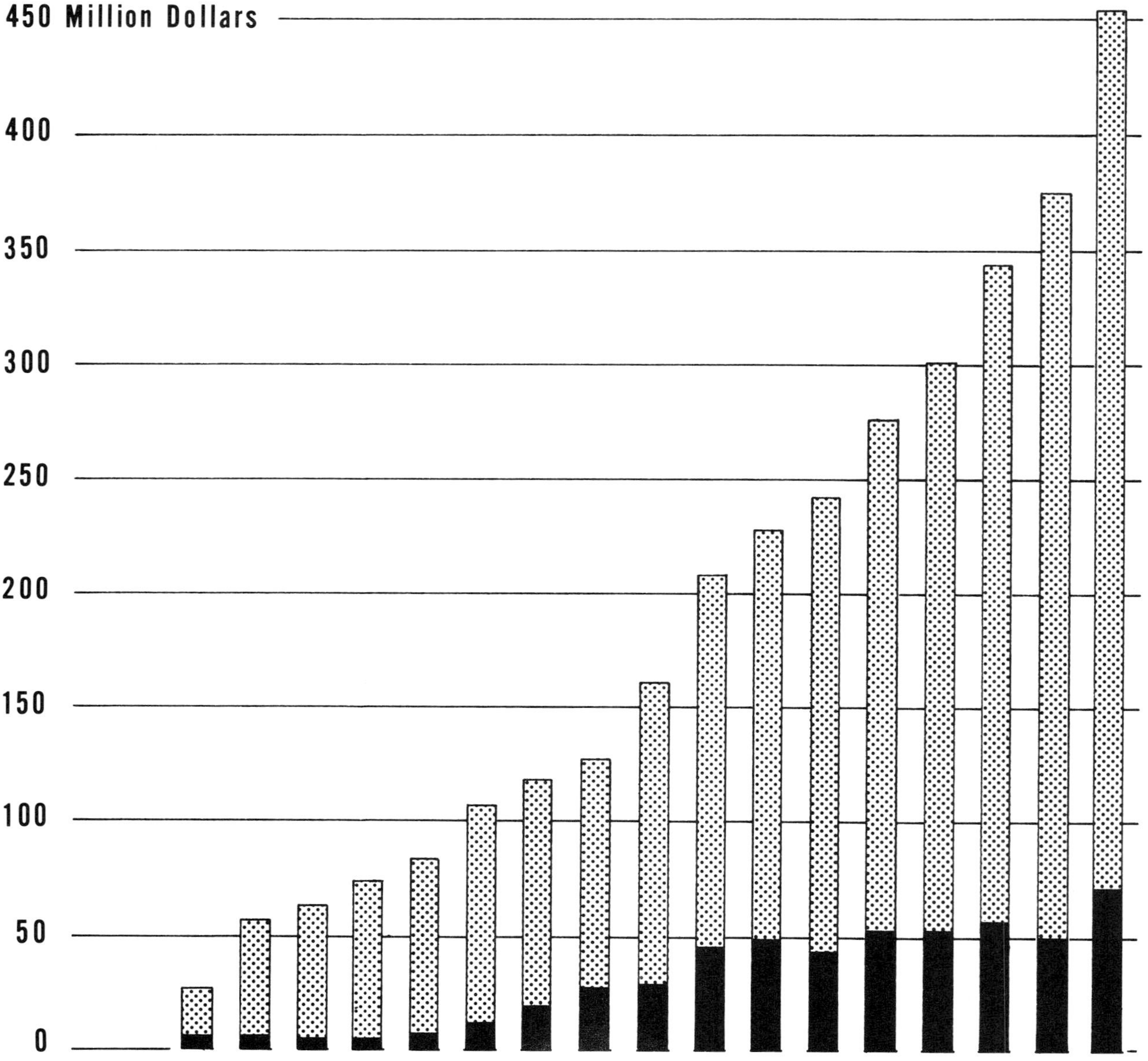

	1950	1952	1954	1956	1958	1960	1962	1964	1966	1968	1970	1972	1974	1976	1978	1980	1982
CIVILIAN	22	49	58	69	77	94	99	98	130	162	180	198	223	248	288	325	382
MILITARY	6	6	5	5	7	12	20	28	30	47	49	45	53	53	57	50	71
TOTAL	28	55	63	74	84	106	119	126	160	209	229	243	276	301	345	375	453

CHAPTER FIVE: *The 1980s: Poised for the Future*

Ready to Move into the Twenty-first Century. . .

At the Headquarters of the Warner Robins Air Logistics Center. . .

. . .the hub for management and repair of major weapon systems for the United States Air Force.

C-141 Starlifter, backbone of the United States strategic airlift fleet.

Ready to Change and Repair...

The F-15 Eagle gets a change for the better on a Maintenance "Speed-line"

A C-141 "Stretch" gets a checkup and a big fix.

Staying ready. . .

. . .with helicopters managed by Warner Robins standing by for rapid deployment worldwide.

. . .with state-of-the-art equipment in the Airborne Warning and Control System (AWACS), . . .

. . .managed and repaired by the Warner Robins ALC.

Keep ready. . .

. . .with a precision fix for subsystems. . .

. . .or a whole system like the F-15.

Ready to put a fix on. . .

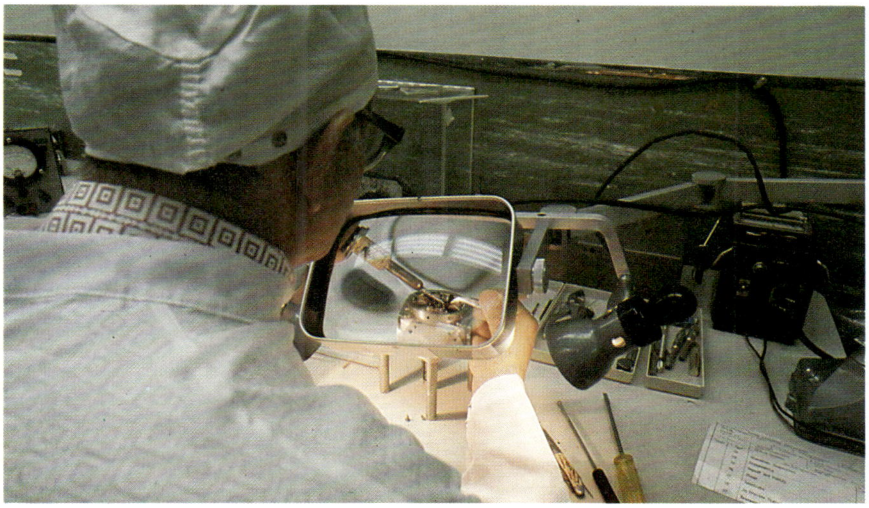

. . .tiny components, . . .

. . .in "clean" rooms. . .

. . .and on big systems out on the line.

Ready with. . .

. . .superb craftsmen and sophisticated equipment. . .

. . .in the Avionics Center of the Air Force.

Ready. . .

. . .with eighteen acres of shops. . .

. . .training centers for new technology. . .

. . .and exotic missions like the skilled test and repair of laser target designators.

Ready with electronic warfare equipment to meet the threat. . .

. . .digital microtechnology at work. . .

. . .a revolution in engineering support,

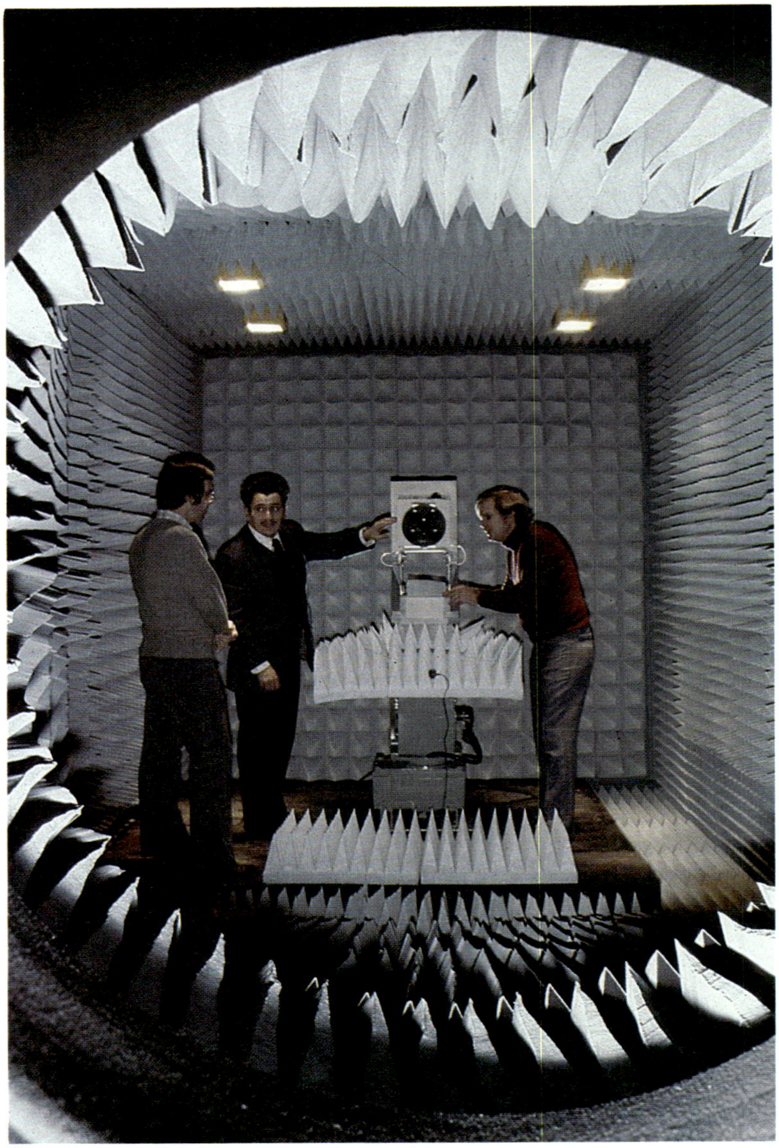

. . .anechoic chambers for engineering checkout of systems to detect, identify and destroy hostile radar. . .

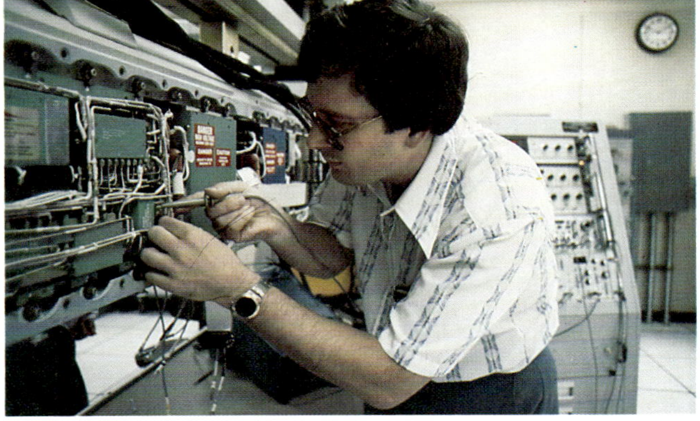

. . .and electronic warfare pods to allow program changes on the flightline without changing hardware.

Ready and working with our community to train engineers and technicians on the job, in the classroom and with private industry. . .

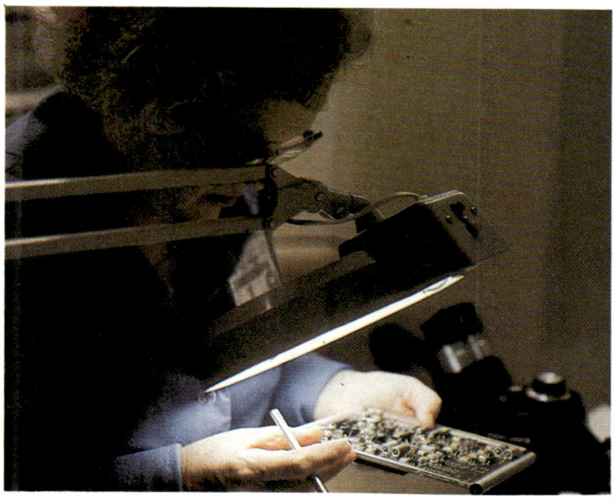

. . .and to meet the defense needs of the United States with a huge land exchange to make room for. . .

. . .the Southeastern PAVE PAWS radar warning system which will watch for enemy sea-launched ballistic missiles.

Ready and working together for national defense. . .

. . .managers, engineers, pilots, financial experts, skilled people and fine facilities.

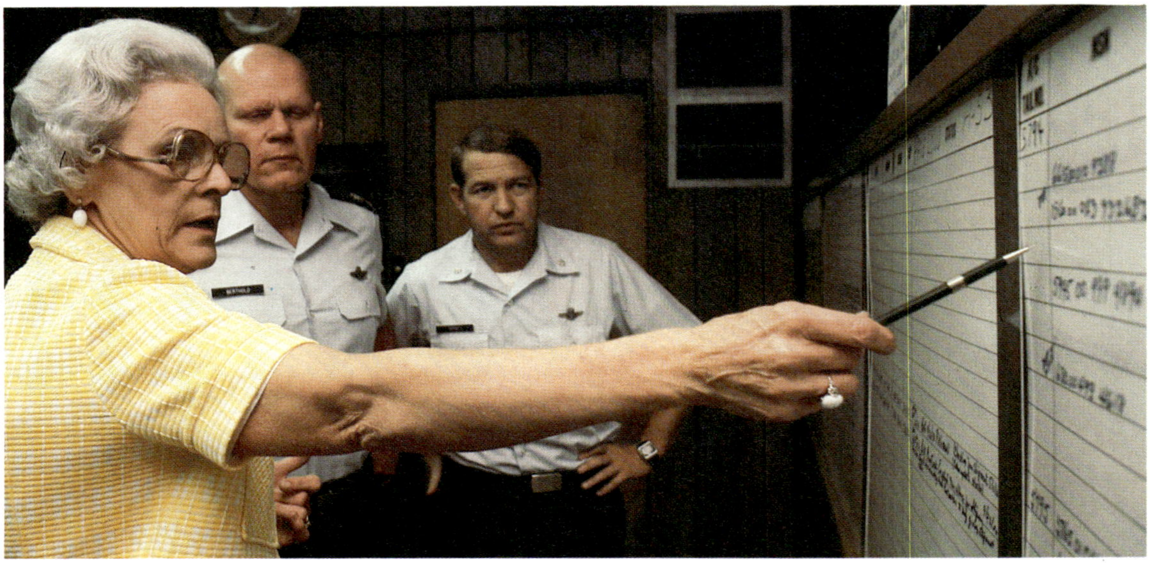

Electronic warfare and software engineering complex.

Ready with computers. . .

. . .testing equipment and programs. . .

. . .flying airplanes. . .

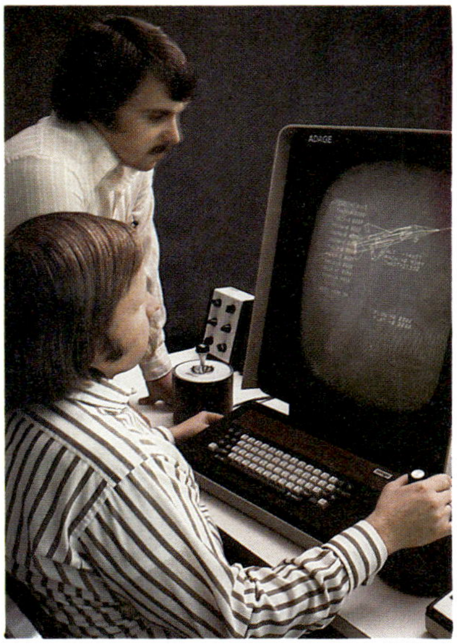

. . .simulating flight. . .

. . .putting it all together for the Air Logistics Center.

Ready with a whole team. . .

. . .contracting to help our allies. . .

. . .moving cargo by air. . .

. . .and on the ground to United States fighting forces worldwide.

Building for the future...

...dormitories for better living...

...the faith of our fathers...

..."energizing" our workload...

...supporting units of other Commands at Robins...

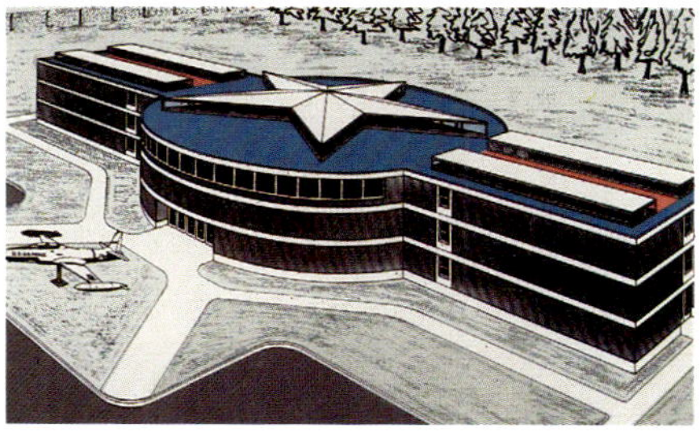

...and remembering our heritage with a proposal for a Robins Air Force Base Museum.

Robins Air Force Base. . .

. . .supporting peace worldwide. . .

. . .working together to protect our American way of life, the land of the free and the home of the brave.

A PHOTOGALLERY: *Leaders of the 1980s*

Commanders, Warner Robins Air Logistics Center

John R. Paulk was born January 9, 1931, in Willacoochee, Georgia, where he graduated from high school in 1948. He attended junior college until 1950 when he enlisted in the U. S. Air Force. Paulk received his pilot wings and commission in December 1952 through the aviation cadet program.

He served as an F-86 fighter pilot with the 18th Fighter-Bomber Wing, in South Korea from March 1953 until April 1954. Following an assignment with the 3505th Pilot Training Wing at Greenville AFB, Mississippi, as an instructor pilot and flight test maintenance officer, Paulk served with the Northern Air Materiel Area, Pacific at Tachikawa Air Base, Japan, as a depot flight test maintenance officer. When he returned from Japan, he was assigned to the 4th Tactical Fighter Wing at Seymour Johnson AFB, North Carolina, as flight test, quality control and maintenance officer, flying F-100s and F-105 Thunderchiefs. In August 1964, Paulk transferred to HQ TAC, Langley AFB, Virginia, as operations staff officer. In January 1965 he was assigned as a senior operations adviser to the 516th Vietnamese Fighter Squadron at Da Nang Air Base, Vietnam, where he flew A-1H Skyraiders. In September of the same year he went to Moody AFB, Georgia, where he served as operations officer and commander of the 3552nd Pilot Training Squadron.

Major General John R. Paulk
January 1980—August 1982

Following a tour at the Military Personnel Center at Randolph AFB, Texas, he returned to Southeast Asia in June 1969 as operations officer and commander of the 435th Tactical Fighter Squadron at Ubon Royal Thai AFB, Thailand, flying F-4s with the 8th Tactical Fighter Wing's "Wolf Pack." In September 1970, Paulk was stationed in England at Royal Air Force Station Bentwaters as deputy commander for maintenance, 81st Tactical Fighter Wing, and was then at RAF Station Lakenheath and served as deputy commander for maintenance, vice commander and later as commander of the 48th Tactical Fighter Wing. He returned to RAF Station Bentwaters in April 1974 as commander of the 81st Tactical Fighter Wing.

General Paulk was named Vice Commander of the Ogden Air Logistics Center, Hill AFB, Utah, in April 1975. In February 1977 he became Deputy Chief of Staff, Logistics, for the North American Air Defense Command and the Aerospace Defense Command. He was named Commander of the Warner Robins Air Logistics Center in February 1980. General Paulk announced his retirement from active duty effective 1 September 1982; future plans were not announced.

General Paulk is married to Gail Council Paulk of Ruskin, Florida. They have a son, Jeff, and a daughter, Eve.

Major General Cornelius Nugteren
September 1982—present

Cornelius Nugteren was born February 7, 1928, in Colton, South Dakota. Following his graduation from high school in Marion, South Dakota, he earned a degree in education from Central College in Pella, Iowa in 1951. He enlisted in the Air Force in 1952 and was commissioned in March 1953 through the aviation cadet program. He won his wings at Williams AFB, Arizona.

General Nugteren has had varied assignments in Germany. In 1953 he flew F-86s amd F-100s with the 36th Tactical Fighter Wing at Bitburg Air Base. In 1960 he was maintenance officer and chief of field maintenance for the 49th Tactical Fighter Wing at Hahn Air Base and in June 1973 became vice commander of the 322nd Tactical Airlift Wing, Rhein Main Air Base. He assumed command of the wing there in February 1974. In August 1975, General Nugteren was assigned to Ramstein Air Base, Germany, as commander of the 86th Tactical Fighter Wing and served as the first Air Force community commander for the Kaiserslautern, Germany, area.

General Nugteren's stateside assignments have included Seymour Johnson AFB, North Carolina; Headquarters, Twelfth Air Force, Waco, Texas; the U. S. Air Force Academy in Colorado; McChord AFB, Washington as a pilot and flight test officer; and two tours at Scott AFB, Illinois, first with Hq Military Airlift Command's Inspector General office and later as commander Aerospace Rescue and Recovery Service. He served as Vice Commander, Ogden Air Logistics Center, Hill AFB, Utah. He has been serving as chief of the Joint United States Military Aid Group in Greece, representing the Secretary of Defense and the U. S. Commander in Chief, Europe. He is to assume command of the Warner Robins Air Logistics Center in September 1982.

General Nugteren is married to Liane Albrecht of Schwelm, Germany. They have two daughters, Cecile, an Air Force lieutenant, and Aneli, who attends Indiana University.

Maj. General Edward Dillon
April 1979—April 1982

Vice Commander, HQ Air Force Reserve

General Dillon was born March 29, 1932 in Santa Monica, California. He graduated from Troy State University, Troy, Alabama. He enlisted in the Air Force in 1951, and his first assignment was at Clovis AFB, New Mexico. He served with the Reserves in California at Long Beach International Airport, March AFB, and Norton AFB. In 1970 he became commander of the 912 Military Airlift Group at Dover AFB, Delaware and in 1973 took command of the 459th Tactical Airlift Wing at Andrews AFB, Maryland. Following his service at Robins, he was reassigned as Mobilization Augmentee to the Vice Commander of NORAD at Colorado Springs, Colorado.

Vice Commanders, Warner Robins ALC

Brig. General Bernard T. Randolph
June 1980—September 1981

Brig. General William M. Shaw, Jr.
October 1981—Present

Biographical Notes on the WR-ALC Vice Commanders:

Brigadier General Randolph: General Randolph was born in New Orleans. Graduating from Xavier University of Louisiana, he completed aviation cadet training at Ellington AFB, TX, and Mather AFB, CA. He then served at Lincoln AFB, NE; at Los Angeles Air Force Station, CA, with the Space Systems Division; in Vietnam as an airlift operations officer at Chu Lai and airlift coordinator at Tan Son Nhut Air Base; with HQ AFSC at Andrews AFB, MD.; returning to Los Angeles Air Force Station as director, space systems planning for the Space and Missile Systems Organization. In April 1975 he became deputy program director and later program director for the Air Force Satellite Communications System. He assumed responsibility for space defense systems at Space Division headquarters, Los Angeles Air Force Station in March 1978. In this capacity he managed a program to design and develop the U.S. anti-satellite system with its supporting surveillance, command and control, and survivability aspects. Following his tour at Robins, Major General Randolph was assigned to his present duties as Director of Space System and Command, Control, Communications, Deputy Chief of Staff, Research and Development, Acquisitions, U.S. Air Force, Washington D.C. in August 1981. General Randolph is married to the former Lucille Robinson of New Orleans. They have six children: Michelle, Julie, Michael, John, Liane and Mark.

Brigadier General Shaw: General Shaw was born April 13, 1934, in Ashville, NC. He received a Bachelor of Industrial Engineering degree and a commission through the ROTC programs from North Carolina State University. He received a MS degree in engineering from St. Mary's University in 1973 and is a registered professional engineer in Texas. He entered the Air Force in 1955 and is a Master Navigator with more than 5,000 flying hours. General Shaw went to Korea in 1968 as Chief of Operations and later deputy commander of a civil engineering squadron. Afterward, he served four years at San Antonio, TX as Chief of Programs and then as assistant to the Base Civil Engineer. His next assignment was at Hill AFB, UT where he was the Base Civil Engineer until July 1976. In 1977 he was Director of Operations and Maintenance and later assistant to the DCS of Engineering and Services at HQ AFLC at Wright-Patterson AFB, OH. Immediately prior to his tour at Robins, he was commander, AFLC Logistics Support Group, Al Khobar, Saudi Arabia. General Shaw is married to the former Alfrieda Ring from Tifton, GA. They have three children—Captian William Martin III, Katherine, and Carol.

WR-ALC Executive Assistant/ Directors/Deputies and Tenant Commanders:

Mr. Lucian M. Rucker
Executive Assistant
August 1981—present

Col. Thomas A. LaPlante
Director of Distribution
September 1980—June 1981

Col. Frederick C. Anderson
Director of Distribution
June 1981—June 1982

Col. Herbert C. Tygett
Director of Distribution
June 1982—present

Mr. Clinton Lewis
Deputy Director of Distribution
January 1981—present

Col. Lee V. Greer
Director of Maintenance
September 1980—April 1982

Col. Richard F. Gillis
Director of Maintenance
April 1982—Present

Col. Thomas A. LaPlante
Director, Materiel Management
June 1981—February 1982

Col. Billy G. Edenfield
Director, Materiel Management
May 1982—present

Mr. Morris E. Ezell
Deputy Director, Materiel Management
June 1980—present

Col. Forrest E. Storz
Director of Plans and Programs
August 1980—Present

Mr. Robert L. Burns
Deputy Director of Plans and Programs
August 1981—Present

Col. Frederick C. Anderson
Base Commander
April 1980—June 1981

Col. Charles E. Pugh
Base Commander
June 1981—November 1981

Col. Leroy W. Thornal
Base Commander
November 1981—Present

Col. Gary H. Bowman
Comptroller
March 1982—Present

Col. Ellie G. Shuler, Jr.
19th Bomb Wing Commander
January 1980—July 1980

Col. Donald L. Marks
19th Bomb Wing Commander
July 1980—August 1981

Col. Richard L. Purdum
19th Bomb Wing Commander
August 1981—Present

Col. John J. Livengood
5th Combat Communications Group Commander
May 1981—Present

Biographical Notes on the WR-ALC Executive Assistant Directors/Deputies and Tenant Commanders:

Mr. Lucian M. Rucker: Mr. Rucker was assigned as the Director of Plans and Programs until he was named as Executive Assistant. He was heavily involved in plans that helped bring the total USAF avionics workload and the F-15 management to Robins.

Col. Thomas A. LaPlante: Colonel LaPlante led in the completion of the Air Freight Terminal and in the selection of Warner Robins as the test site for a new system called Logistics Application of Automated Marking and Reading Symbols (LOGMARS). During his tenure, the Directorate received the third rating of "Excellent" from the AFLC Inspector General. He left the Directorate to become the Director of Materiel Management at WR-ALC and subsequently was promoted to Brigadier General as Vice Commander, Air Force Acquisition Logistics Division, Headquarters Air Force Logistics Command.

Col. Frederick C. Anderson: Colonel Anderson led in the successful program to achieve one million dollars in savings in the Value Engineering Program in the Directorate, completing twelve projects with validated savings of $1.6 million, the first group on base to reach the goal. He directed the establishment of model programs of Affirmative Action and in the Federal Employment Opportunity Recruitment Program (FEORP). He retired in June 1982 and moved to Revloc, PA.

Col. Herbert C. Tygett: Colonel Tygett was previously assigned as Chief of the Resources Management Division, Directorate of Materiel Management at Robins.

Mr. Clinton Lewis: Mr. Lewis came to the Directorate from the Directorate of Maintenance. He has continued programs to enhance the meaningfulness of employment within the Directorate, the sense of mission identification and excellence of performance.

Col. Lee V. Greer: Under his direction, the *Have Swap* Program resulting from a shortage of F-15 engines was completed. The thirty F-15s used were processed through Speedline for latest modifications and returned to TAC ahead of schedule. He established monthly Union-Managment meetings to further improve relations between management and the union. A native of Georgia, Colonel Greer was the first Director of Maintenance to be selected for the rank of Brigadier General. He was reassigned as Vice Commander, Sacramento Air Logistics Center, CA.

Col. Richard F. Gillis: Colonel Gillis came to Robins from AFALD at Wright-Patterson AFB, OH, where he served as the Director, Logistics for Airlift and Training Systems.

Col. Thomas A. LaPlante: Brigadier General LaPlante transferred from his position as Director of Materiel Management and is serving as Vice Commander, Air Force Acquisition Logistics Division, Air Force Logistics Command.

Col. Billy G. Edenfield: Colonel Edenfield previously served at the rank of Major in the Directorate as logistics systems manager for the C-130, VC-140 and B-57 aircraft.

Mr. Morris E. Ezell: Mr. Ezell, a member of the Senior Executive Service, transferred to his present position from Wright-Patterson Air Force Base, OH, where he was Assistant to the Commander, Air Force Acquisition Logistics Division.

Col. Forrest E. Storz: Colonel Storz was formerly assigned as chief of the Resources Management Divison in the WR-ALC Directorate of Maintenance. He came to Robins from Wright-Patterson AFB, OH, where he was Director of Programs at HQ AFLC.

Mr. Robert L. Burns: Mr. Burns was assigned as Executive Assistant to the ALC Commander prior to accepting the Deputy Directorship.

Col. Charles E. Pugh: Colonel Pugh transferred to WR-ALC from HQ ADCOM. Colonel Pugh retired from active duty at Robins.

Col. Leroy W. Thornal: Col. Thornal was previously assigned as RAFB Civil Engineer.

Col. Gary H. Bowman: Col. Bowman came to WR-ALC from HQ MAC where he served as Director of Industrial Funds. He had previously served as Chief of the Budget Management Division, HQ AFLC.

Col. Ellie G. Shuler, Jr.: Colonel Shuler left the wing to become commander of the 42nd Bomb Wing at Loring AFB, Maine. He was promoted to Brigadier General and now serves as Commander, 4th Air Division, Francis E. Warren AFB, WY.

Col. Donald L. Marks: During his tenure the 19th Organizational Maintenance Squadron received the Outstanding Unit Award. Also, during the Operational Readiness Inspection in April 1981, the wing achieved an excellent rating. Colonel Marks departed Robins to become the Wing Commander for the 42nd Bomb Wing, at Loring AFB, ME.

Col. Richard L. Purdum: During his tenure the wing has achieved many outstanding awards. The wing placed first in the 1981 Giant Sword Loading Competition. It was selected as the "Best in SAC" in Aircraft Maintenance and was the runner-up for the Air Force Daedalian Maintenance Award. The wing received the General George C. Kenney Award for attaining the highest bombing scores in an Operational Readiness Inspection (ORI), the score of .98 damage expectancy (CE) was the highest ever attained in the command. In February 1982, the wing received the Warner Robins-Houston County Chamber of Commerce Good Government Service Award. In May 1982, the wing was awarded the Omaha Trophy as the most outstanding wing in SAC for 1981.

Col. John J. Livengood: Under his leadership the group has been designated the lead combat communications group supporting the Rapid Deployment Force (RDF) and has been selected as the lead unit to convert from the BAMS to the COMPS mobility system. He was on hand to receive the first new Quick Reaction Package (QRP) in March 1982.

Affiliated Units Sharing Robins Air Force Base and Their Current Commanders

Air Force Reserves (AFRES)	Vacant
19th Bombardment Wing (SAC)	Col. Richard L. Purdum
5th Combat Communications Group	Col. John J. Livengood
Det 5, AF Communications Command	Col. John G. Nash
1926th Communications Squadron	Lt. Col. Abraham Bronstein
1839th Engineering Installation Group	Mr. Fleming G. Gilman
3503rd USAF Recruiting Group	Col. David W. Saunders
321st Field Training Detachment	Capt. Deborah M. Brown
14th Flying Training Wing Detachment (ACE)	Capt. Paul O. Decker
Det 8, 2762nd Logistics Squadron (Special)	Col. Richard Scuderi
AFLC NCO Academy Leadership School	CMSgt. Larry D. Huyett
Det 6, 3025 Management Engineering Team	Lt. Col. William H. McDaniel
Det 3, 2 Aircraft Delivery Group	Lt. Col. Thomas E. Gassman
4400th Mobility Support Flight	Maj. Richard K. Sutton
Det 13, 15th Weather Squadron	Lt. Col. Thomas E. Kretlow
Base Commissary	Mr. Eugene T. Hamlin
Area Defense Council	Capt. Bradley S. Adams
Air Force Office of Special Investigations, Det 712	Special Agent Ronald K. Bliss
Air Force Audit Agency Area Office, Det 960	Lt. Col. Larry E. Faith
Defense Property Disposal Office	Mr. Clifford N. Cline
Defense Contract Administration Services, Atlanta Management Region, Branch 2	Mr. Marvin Evans
DOD Placement Coordinator Office	Mr. Lee C. Lingelbach
Defense Investigative Service (District 41)	Mr. John H. Bryant
Base Exchange	Mr. Kenneth A. Grimes
US Army Corp of Engineers	Mr. Elton L. Cobb
Macon Radar Approach Control/Tower	Mr. Lloyd K. Alley
Small Business Administration	Mr. James C. Parker
US General Accounting Office	Mr. Robert F. Rehberg
C-130 Liaison Office Royal Air Force	Squadron Leader Henry R. Smith
1 Canadian Forces Logistics Liaison Unit	Capt. H. Elmer Kellett
Royal Australian AF Technical Liaison Office	Squadron Leader Gerald A. Roberts
Israeli Air Force Liaison Office	Lt. Col. Oved Evry
F-15 Technical Liaison Officer, Japanese Air Service Defense Force	Maj. Hide Asano
Saudi Arabia Air Force Liaison Officer	Capt. Samah Moed Al-Anazy

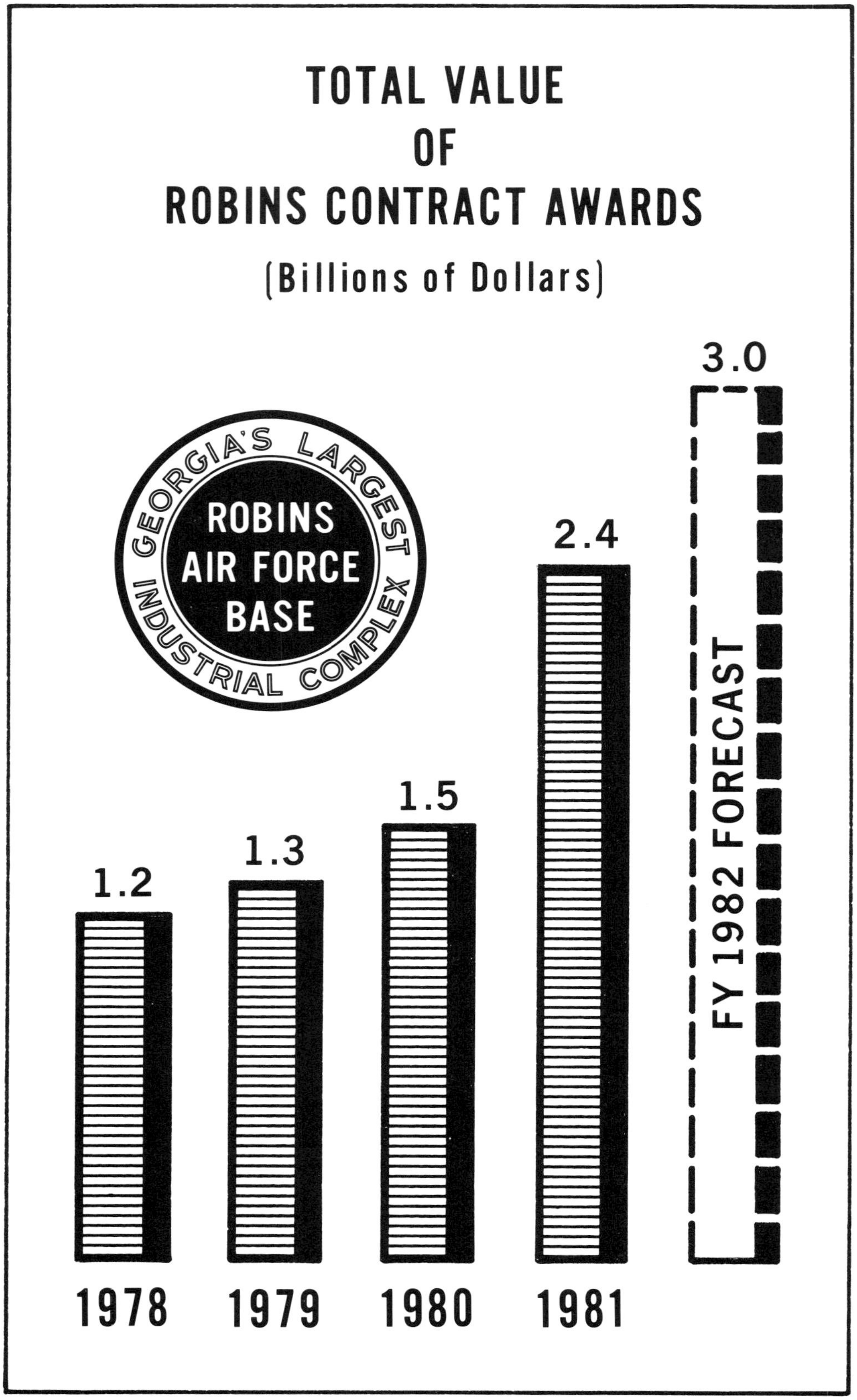
TOTAL VALUE
OF
ROBINS CONTRACT AWARDS
(Billions of Dollars)
GEORGIA'S LARGEST INDUSTRIAL COMPLEX
ROBINS AIR FORCE BASE
1.2
1.3
1.5
2.4
3.0
FY 1982 FORECAST
1978
1979
1980
1981

Epilogue

The history of Robins Air Force Base is that of a military installation hurriedly built to meet a tardily perceived threat to American freedom.

Remote from any large city, in what was the most industrially undeveloped section of the nation, it grew into the technologically advanced giant which now calls itself the "Avionics Center of the Air Force." It owes much to a long succession of able and dedicated leaders, both military and civilian. It owes no less to an endless line of men and women who believed they should give a day's work for a day's pay and that love of country was more than rousing speeches and fireworks on the Fourth of July.

Colonel Charles E. Thomas, Jr., at the dedication in 1943 of what is now the Warner Robins Air Logistics Center correctly predicted the future of the depot up to the present time. His words are as fresh in 1982 as when he spoke them 39 years ago:

> The depot shall be here keeping our Air Forces flying long after the victory. We are building with that in mind—that you and we are permanent neighbors . . . Together we are moving toward the better days of which we now dream . . . and, together, when those days come, we will help perpetuate them by continuing our support of the greatest flying forces in all the world.

If America could be made forever secure by a single generation of Americans, those who have gone before us would have done it. Unfortunately, eternal freedom requires eternal vigilance. It is the least price that every generation must pay and many have paid a great deal more. If the history of Robins Air Force Base has any value, it is in whatever insight and inspiration it might provide to present and future Robins personnel. As Carl Vinson said shortly before his death:

> I am convinced that what is past is prologue. I hope that this generation and the generations to follow will be guided by this warning from an old man—weakness and indecision invite disaster!

List of selected land grants, 5th and 11th District, Houston County, being the land lots on which Robins Air Force Base Now Stands

Source: Surveyor General's Department, State Archives, Atlanta

Landlot	*Grantee*	*County of Residence*
214	Thomas White	Jones
215	James Ennis	Camden
216	Joseph L. Barnett	Gwinnett
217	Elizabeth Legett (widow)	Oglethorpe
218	Sarah Barnett	Washington
219	Jonas Haite	Clarke
224	Drury A. McCullens	Twiggs
225	William Vines (orphan)	Jones
226	John Fudge	Richmond
227	James Williams	Tattnall
228	Thomas Rousseau	Baldwin
229	Cullen Davis	Hall
230	Mary Dews (widow)	Chatham
235	Elizabeth Cato (widow)	Emanuel
236	Washington Glover	Warren
237	Whitmill Hardy	Hancock
238	Cullen Boyett	Richmond
239	Morgan Ward	Liberty
240	Jonathan Dawson	Jones
241	Edmind Palmer	Richmond
242	Benjamin Smith	Pulaski
243	Uliat Lee	Clark
244	Bartholomew Murphey	Pulaski
245	William C. Thompson	Franklin
246	Robert Smith	Laurens
247	Dennis Dodd	Elbert
248	Jesse Prossor	Baldwin
249	Daniel Beall	Morgan
250	Nathan Minshew	Twiggs
255	Phillip Combs, Jr.	Wilkes
256	John Coldwell	Richmond
257	Lewis Tyar	Appling
258	Eleanor Davis	Columbia
259	Sidney Liverman	Richmond
260	Joseph Edwards	Franklin
262	Charles McDaniel	Washington
263	Thomas Edmonson	Putnam
264	Georgianna Miller	Lincoln
265	Moses Overstreet	Burke
266	Elijah Terry	Morgan
47	Theophilus Oneal	Laurens
48	Allison Kent	Morgan
49	John Platt	Bulloch
80	John M. Shivery	Washington

The last four land lots are in the 11th district; others are in the 5th district.

Major buildings constructed on Robins Air Force Base
(whose current cost exceeds $100,000)

Building Number	*Civil Engineering Nomenclature*	*Square Feet*	*Current cost*	*Year First Completed*
1	Electric Power Station	2,039	$650,577.55	1980
2	RAPCON Center	6,036	$292,325.14	1960
8	Operations, Ordnance Control Point	2,624	$103,487.63	1972
12	Readiness, Crew	19,057	$712,397.25	1960
38	Hydrant Fuel Building	942	$141,848.19	1967
43	NCO Professional Education Center	8,678	$209,237.97	1959
44	Maintenance Dock, Large Aircraft	12,694	$293,000.69	1956
45	Maintenance Dock, Small Aircraft	7,554	$102,294.89	1955
46	Communications Facility	5,700	$109,095.14	1969
47	Maintenance Dock, Large Aircraft	26,790	$623,838.32	1970
48	Maintenance Dock, Large Aircraft	26,790	$538,849.69	1967
49	Maintenance Dock, Large Aircraft	26,790	$538,706.22	1967
52	Maintenance Dock, Large Aircraft	17,740	$343,546.87	1962
53	Compressed Air Plant Building	703	$102,483.75	1959
54	Aircraft Corrosion Control	53,029	$3,695,421.87	1978
55	Maintenance Dock, Large Aircraft	15,694	$254,757.47	1956
56	Headquarters, Group	10,439	$223,862.03	1959
59	Warehouse, Supply and Equipment Base	43,667	$257,542.66	1961
66	Maintenance Dock, Large Aircraft	30,897	$403,217.46	1960
67	Maintenance Dock, Large Aircraft	20,463	$244,779.61	1960
74	Electric Power Station	1,710	$567,793.53	1980
76	Shop, A/C Maint. Eng. Inspect/Repair	33,238	$526,615.25	1960
78	Headquarters, Wing	35,039	$732,988.45	1960
79	Shop, Aircraft General Purpose	53,737	$1,110,361.52	1960
80	Security Police Central Control	4,894	$121,826.46	1961
81	Maintenance Dock, Large Aircraft	20,463	$299,710.61	1960
82	Shop, Aerospace Support Equip. Storage	21,183	$366,568.59	1960
86	Shop, Aerospace Support Equip. Storage	4,768	$122,190.64	1960
89	Hangar, Maintenance Depot	56,954	$3,583,439.44	1967
95	Water Fire Pump Station	1,597	$142,237.03	1956
97	Storage, Multicubical Magazine	7,776	$118,297.93	1960
99	Storage, Igloo	6,992	$226,023.93	1974
100	Shop, Missile Assembly	8,268	$611,802.79	1974
107	Security Police Entry Control Building	2,207	$188,871.40	1960
109	Fire Station	9,326	$147,197.24	1956
110	Aircraft Corrosion Control	160,668	$3,364,935.08	1942
111	Operations, Base	3,690	$110,931.98	1963
125	Hangar, Maintenance Depot	593,291	$11,121,335.28	1942
127	Terminal, Air Freight	67,782	$1,961,907.07	1978
128	Shop, Survival Equipment	14,385	$321,144.59	1953
140	Shop, Aircraft General Purpose	220,105	$4,115,511.29	1974
142	Shop, Aircraft General Purpose	44,972	$1,161,826.70	1970
148	Shop, Aerospace Support Equip. Storage	43,960	$545,067.70	1944
149	Hangar, Maintenance Depot	44,240	$502,760.20	1954
155	Logistical Facility, Depot Operations	24,726	$389,986.70	1944
158	Shop, Instrument Overhaul Depot	71,184	$1,992,141.99	1942
159	Fire Station	3,000	$101,393.64	1981
162	Laboratory, Precision Measure. Equip.	44,818	$1,251,588.02	1942
165	Laboratory, Quality Control, Depot	30,677	$1,196,184.13	1942
166	Base Restaurant	38,874	$697,067.74	1942
169	Shop, Aircraft General Purpose	171,781	$3,104,143.92	1942
173	Shop, A/C Maintenance, Organizational	40,000	$247,776.89	1942
177	Steam Facility	24,310	$4,435,279.17	1942

Building Number	*Civil Engineering Nomenclature*	*Square Feet*	*Current cost*	*Year First Completed*
181	Shop, Aircraft General Purpose	69,408	$3,404,168.90	1942
183	Logistical Facility, Depot Operations	18,016	$163,623.18	1943
184	Club, Aero	4,212	$113,111.36	1980
190	Shop, refueling vehicle	3,575	$124,887.28	1963
210	Headquarters, Major Command	113,598	$2,463,961.45	1955
214	Communications facility	14,517	$303,971.84	1959
215	Headquarters, Group	89,100	$1,162,435.61	1955
220	Logistical Facility, Depot Operations	58,082	$628,343.93	1942
225	Communications facility	14,869	$692,188.05	1957
226	Integrated support facility	126,627	$5,438,149.33	1980
229	Integrated support facility	43,907	$636,746.19	1943
230	Warehouse, Forms/Publications, Base	40,000	$120,146.19	1943
234	Occupational Medicine Services	9,000	$119,531.84	1942
250	Logistical Facility, Depot Operations	10,745	$118,330.50	1942
263	Security Police Operations	8,937	$201,425.03	1968
272	Base Engineer Maintenance Shop	15,742	$455,202.35	1943
274	Family Housing Management Office	7,555	$194,953.08	1943
280	Base Engineer Administration	12,033	$183,714.70	1955
282	Base Engineer Storage, Covered	30,883	$270,728.07	1955
300	Logistical Facility, Depot Operations	440,348	$6,769,830.08	1942
301	Warehouse, Support Equipment Depot	428,177	$2,860,719.63	1942
302	Vehicle Maintenance Shop	22,374	$163,441.06	1942
307	Vehicle Maintenance Shop	12,034	$128,585.80	1942
309	Warehouse, Support Equipment Depot	156,625	$666,825.43	1942
310	Warehouse, Support Equipment Depot	156,625	$389,588.53	1942
321	Plant, Printing	24,375	$268,963.98	1944
323	Technical Laboratory	19,764	$153,981.96	1955
327	Hazardous Storage, Base	21,867	$178,806.70	1942
332	Warehouse, Support Equipment Depot	108,000	$186,400.00	1968
350	Warehouse, Support Equipment Depot	244,417	$1,249,129.21	1957
351	Material Processing, Depot	120,767	$767,273.06	1955
364	Warehouse, Support Equipment Depot	152,308	$476,808.85	1943
365	Material Processing, Depot	152,308	$448,139.48	1943
366	Warehouse, Support Equipment Depot	152,308	$315,361.73	1943
367	Warehouse, Support Equipment Depot	152,308	$249,790.47	1943
368	Warehouse, Support Equipment Depot	324,417	$1,569,533.30	1957
376	Material Processing, Depot	400,000	$5,701,271.01	1975
377	Fire Station	9,864	$427,623.10	1977
380	Warehouse, Support Equipment Depot	573,512	$3,180,768.92	1956
385	Warehouse, Support Equipment Depot	651,979	$3,955,418.07	1956
542	Open Mess, Officers	25,463	$295,371.28	1942
551	Visiting Officers Quarters	17,150	$270,561.44	1960
552	Visiting Officers Quarters	17,150	$247,083.28	1960
553	Dormitory, Visiting Airmens Quarters	15,726	$376,070.71	1969
557	Visiting Officers Quarters	17,150	$256,617.38	1960
595	Golf Clubhouse/Equipment	7,278	$280,872.52	1977
603	Shop, Aircraft General Purpose	20,800	$121,498.96	1942
608	AFCS Maintenance Facility	20,257	$134,043.83	1942
614	AFCS Maintenance Facility	12,476	$223,141.01	1960
628	Air Conditioning Plant Building	2,727	$146,882.11	1954
640	Shop, Avionics	264,855	$4,592,819.18	1954
641	Warehouse, Support Equipment Depot	324,148	$3,326,003.08	1974
642	Air Conditioning Plant Building	4,994	$931,211.27	1974
644	Steam Facility Building	5,693	$538,343.57	1953
645	Shop, Avionics	201,335	$5,569,115.87	1953
653	Base Engineer Covered Storage	12,600	$131,335.23	1942
660	Warehouse, Support Equipment Depot	250,724	$2,262,207.19	1957
670	Shop, Aircraft General Purpose	33,703	$474,548.23	1944
700	Composite Medical	74,112	$2,500,106.75	1967
755	Dormitory, Airmen Permanent Party	24,961	$356,656.56	1957

Building Number	*Civil Engineering Nomenclature*	*Square Feet*	*Current cost*	*Year First Completed*
756	Dormitory, Airmen Permanent Party	24,961	$327,345.50	1957
757	Dining Hall	14,513	$445,507.83	1960
758	Dormitory, Airmen Permanent Party	25,000	$338,105.36	1960
759	Dormitory, Airmen Permanent Party	25,000	$304,516.63	1960
760	Band Center	5,446	$154,046.25	1943
761	Field Training Facility	10,788	$211,268.05	1972
767	Recreation Center	27,775	$417,831.51	1959
769	Chapel Center	16,883	$526,597.22	1963
780	Dormitory, Airmen Permanent Party	25,000	$304,857.93	1960
781	Dormitory, Airmen Permanent Party	25,000	$302,779.68	1960
794	Headquarters, Group	8,059	$279,861.07	1961
795	Dormitory, Airmen Permanent Party	25,129	$424,747.22	1968
796	Dormitory, Airmen Permanent Party	25,129	$454,608.52	1968
797	Dormitory, Airmen Permanent Party	25,346	$424,893.85	1969
798	Dormitory, Airmen Permanent Party	25,346	$430,883.72	1969
827	Gymnasium	24,546	$439,885.07	1960
901	Technical Training Classroom	6,793	$148,899.52	1960
908	Bowl Center	13,809	$217,854.96	1964
909	Theater, Base	12,067	$386,379.00	1974
910	Post Office Center	5,150	$140,474.85	1972
914	Exchange, Sales Store	39,866	$821,946.57	1971
915	Cold Storage, Base	5,494	$230,461.34	1972
922	Exchange, Service Station	3,375	$201,967.00	1974
937	Base Personnel Office	6,620	$107,705.91	1961
946	Child Care Center	7,100	$193,242.45	1975
948	Headquarters, Group	16,420	$309,996.80	1970
949	AFCS Maintenance Facility	8,783	$236,881.82	1967
950	AFCS Maintenance Facility	4,255	$118,312.13	1967
956	Open Mess, Non-Commissioned Officers	23,892	$840,127.38	1975
984	Arts and Crafts Center	7,289	$181,200.86	1971
1021	Youth Center	11,676	$468,704.29	1979
1180	Temporary Lodging Facility	4,451	$106,140.48	1972
1181	Temporary Lodging Facility	4,451	$106,140.48	1972
1182	Temporary Lodging Facility	4,451	$106,140.48	1972
1183	Temporary Lodging Facility	4,451	$106,140.48	1972
1675	Technical Training Classroom	20,110	$208,927.78	1943
1676	Logistical Facility, Depot Operations	20,110	$169,148.77	1943
1677	Headquarters, Major Command	20,110	$155,976.70	1943
1678	Headquarters, Major Command	20,110	$182,027.25	1943
1679	Administrative Office, Non-Air Force	20,110	$144,243.99	1943
1682	Administrative Office, Non-Air Force	20,110	$164,210.38	1943
1683	Administrative Office, Non-Air Force	20,110	$186,984.45	1943
1686	Technical Training Classroom	20,110	$144,446.15	1943
3823	Building, Water Supply	608	$116,176.33	1977

Bibliography

1. *Histories*

Cochran Field Histories (Extracts from Originals at Air Force Historical Archives, Maxwell AFB), Vol. I, From Activation to 7 December 1941.

A Brief History of Cochran Field, Georgia, 1940-1947.

Completion Report on Construction of Robins Field (Warner Robins Army Air Depot), Job A 42, Wellston, Georgia.

Warner Robins Air Service Command Histories, 1941-1943, 1 July-1 October 1944, 1941-1944, 1 January-1July 1944.

Warner Robins Air Technical Service Command Histories, 1 October-31 December 1944, 1 January-31 March 1945, 1 April-30 June 1945, 1 October-31 December 1945, 1 January-31 March 1946.

Warner Robins Air Materiel Area Histories, 1 July-30 September 1946, 1 October-31 December 1946, January-June 1947, July-December 1947, 1 January-30 June 1948, July-December 1948, January-June 1949, July-December 1949, 1 January-30 June 1950, 1 July-31 December 1950, 1 January-30 June 1951, 1 July-31 December 1951, 1 January-30 June 1952, 1 July-31 December 1952, 1 January-30 June 1953, 1 July-31 December 1953, 1 January-30 June 1954, 1 July-31 December 1954, 1 January-30 June 1955, 1 July-31 December 1955, 1 January-30 June 1956, 1 July-31 December 1956, 1 January-30 June 1957, 1 July-31 December 1957, 1 January-30 June 1958, 1 July-31 December 1958, 1 January-30 June 1959, 1 July 1959-30 June 1960, 1 July 1960-30 June 1961, 1 July 1962-30 June 1963, 1 July 1963-30 June 1964, 1 July 1965-30 June 1966, 1 July 1966-30 June 1967, 1 July 1967-30 June 1968, 1 July 1968-30 June 1969, 1 July 1969-30 June 1970, 1 July 1971-30 June 1972, 1 July 1972-30 June 1973, 1 July 1973-30 June 1974.

The Warner Robins Air Materiel Report for 1964.

Warner Robins Air Materiel Area's Support of the C-130 Aircraft in Southeast Asia, 1965-1969, A Corona Harvest Input.

Bibliography of WRAMA Weapon Systems, July 1969.

Southeast Asia Document Collection, Vol III: WRAMA Aircraft Weapon Systems in Southeast Asia, 1 January 1965-31 Mar 1968 and Vol XI: AC-119G/K Gunship 1967-1969.

EEO Plan of Action, CY74, Report of Accomplishments, WR-ALC.

Warner Robins Air Logistics Center Histories, 1 July 1973-30 June 1974, 1 July 1974-31 December 1975, 1 January 1976-31 December 1976, 1 January 1977-31 December 1977, 1 January 1978-31 December 1978, 1 October 1978-30 September 1979, 1 October 1979-30 September 1980.

A Short History of the Fourteenth Air Force, Robins AFB, 1943-1959.

Major History of HQ Ground Electronics Engineering and Installation Area (*Geeia*), Rome AMA, New York, 1 July-31 December 1960, and 1 January-30 June 1961.

Histories of the Seventh Logistics Support Squadron, Robins AFB, Georgia, 18 October 1954-6 February 1955, 2 January 1960-31 December 1961, 1 January 1962-Deactivation.

News Releases, WRAMA, Office of Information, 1956-1957 and 1958-1960.

History of the AAF Medical Service Training School at Robins Field, Georgia, 1941-1945.

History of the 1926th Communications Squadron, Robins AFB, Georgia, CY 1975.

Post Regulations, Robins Field, Georgia, Revised March 1946.

Macon, Georgia, 1940, Correspondence used in submitting this area for consideration.

Manpower Allotment and Fiscal Project Breakdown (WRAMA), September 1948. Robins Air Force Base, Georgia, 1948.

Supplement to Warner Robins Air Materiel History, 1949-1952.

Robins Air Force Base Story, 1955, 1958.

Draft (handwritten) of Historical Study on Warner Robins Air Materiel Area's Role in the TM-61 Program, 1953-1956 by Dr. Lewis J. Carey.

Warner Robins Air Materiel Area Speeches, 1954-1960, January 1969.

The Cuban Crisis, 1962, Volume II, FY63, Cuban News Clippings; Directives, Feeder Reports, Memos, Notes, etc. 1962; Cuban Articles, Part III; Chronology of the Cuban Crisis as Viewed by WRAMA, Vol I.

Military Construction Program, FY62.

Selected Statements of President Kennedy on Defense Topics, December 1957-1 August 1962.

General Hefley's Speeches, 1960-1963, January 1969.

Housing Documents, 1962-1963.

WRAMA History, Civil Rights Documents, Vol II, III, IV, V & VI.

General Hedlund's Speeches 1963-1966, to Georgia Legislature, 1964.

WRAMA History, Speeches by the Commander and Key Officials of WRAMA, April 1962-July 1968, January 1964-March 1966, August 1966-December 1967, 1971-1973, 1966-1976.

WRAMA Directorate of Maintenance, Significant Southeast Asia Work Requirements, Documents, February 1966-September 1974.

WRAMA History, AC-47 Gunship Documents, 1964-1966, 1965-1969, 1969.

WRAMA History, Project Red Horse, 1966-1967.

Access Roads for Robins Air Force Base, June 1967.

WRAMA Community Relations, 1967-1968.

General Gideon's Speeches 1966-1969.

Project Corona Harvest, Operating Instructions, 24 May 1968.

AC-119G/K Gunship Documents, 1964-1969.

History of WRAMA, Southeast Asia Document Collection, Vol XX: C-130 Aircraft, 1965-1967.

History of WRAMA, Speeches by Major General A. J. Beck, WRAMA Commander, 1968-1969; 1969-1972; 1971; 1971-1972; Vol II, 1970.

Macon, Georgia, Spring Cleaning, 1969; WRAMA/RAFB Voluntary Support, 10 May 1969.

WRAMA Phase-Up Data, 1964-1969, Vol I, II, III; 1971 Vol II; 1965-1967 Schedule (Photographs).

History of WRAMA, Speeches by Brig. Gen. Ralph T. Holland, 1969-1975.

Commanders Digest, DOD, Washington D. C., September 1972-January 1978.

WRAMA Logistics Role, FY57-FY76.

History of WR-ALC, CY76, EEO Plan of Action.

History of WRALC, F-15 Support Posture Conference, 14 November 1978.

AFLC Support of Forces in Southeast Asia: Cargo Aircraft Support, 1965-1968.

History of AFLC, 1917-1973.

2. *Historical Studies*

Historical Study No. 1, The Campaign for more Adequate Housing, FY63, Part 1.

Historical Study No. 1, Vol 1, The Housing Campaign, 1962, FY63.

Historical Study No. 2, The Cuban Crisis, 1962, FY63.

Historical Study No. 3, Civil Rights, FY64.

Historical Study No. 6, (WRAMA) 1 July 1964-30 June 1965, (FY65), Mission Activities, Parts I and II.

Historical Study No. 7, 1 July 1965-30 June 1966, WRAMA's Role in Southeast Asia, Project RED HORSE, 17 October 1965-31 December 1966.

Historical Study No. 8, WRAMA's Role in Southeast Asia, Project *Bitterwine*, 1 May 1965-31 December 1966.

Historical Study No. 9, WRAMA's Role in Southeast Asia, A General Examination, 1 January 1961-31 December 1966.

Historical Study No. 12, 1 July 1967-30 June 1968, CY67, WRAMA's Role in Southeast Asia, Management Aspects

Historical Study No. 16, WRAMA's Role in Southeast Asia, Armor Plating Flare Box, 1967-1968, C-47 Aircraft.

Historical Study No. 17, 1 July 1969-30 June 1970, WRAMA's Role in Southeast Asia, AC-47 Gunship, 1960-1968 (Project Big Shoot).

Historical Study No. 18, 1 July 1969-30 June 1970, Part III: WRAMA's Role in Southeast Asia, AC-119G/K Gunship Program, 1967-1970, Project Combat Hornet.

Historical Study No. 19, WRAMA, Impact of Social, Political, Economic Forces.

Historical Study No. 19, 1 July 1969-30 June 1970, Part IV: A Commentary of the Homefront During the Vietnam Conflict, The Impact of Social, Political and Economic Forces on the Armed Services.

Historical Study No. 25, The AC-130E Aircraft (Gunship) (Project *Pave Spectre*), 1970-1971, Vol I.

Historical Study No. 26, C-133 Cargomaster, 1953-1971.

Historical Study No. 35, History of WR-ALC, FY74, Volume XX, News Releases in FY74.

3. *Periodicals*

"Cuba: Rumors of Russians"; *Newsweek*, September 3, 1962, page 41.

"Heraldry and Your Family Name - Houstoun"; *Georgia Journal*, August/September 1982, page 27.

"Industry Fires Away at Fixed-Price Contracts"; *Business Week*, November 16, 1968.

"Missile Sites in Cuba: Offensive Purposes Against the Hemisphere"; Speech of Dean Rusk delivered to Council of the Organization of American States, October 23, 1962; *Vital Speeches*, November 15, 1962, VOL XXIX, No. 3.

"Our Changing Military Policy: Qreater Flexibility"; *Vital Speeches*, VOL XXVIII, No. 11, March 15, 1962.

"Washington in Crisis"; *Look*, December 18, 1962, pages 43-44, 49-50 and 54.

4. *Newspaper Articles*

"Aged C-47 is Rejuvenated for Ground Attack in Vietnam"; *Washington Post*, November 24, 1965.

"Macon To Be Considered for $14,000,000 Air Depot"; *The Macon Telegraph*, March 7, 1941.

"No Major Logistical Problem"; *Christian Science Monitor*, December 27, 1962

"Robins Cleans Up After Tornado"; *The Robins Times*, May 12, 1953.

"Tornado Victims Tell Stories of Grief, Fear"; *The Macon Telegraph*, May 1, 1953.

"Tribute Paid Gen. Thomas", *The Macon Telegraph*, January 22, 1955.

"U.S. Mustered Force of 340,000 in Crisis"; *The Macon Telegraph*, November 30, 1962.

"Wellston Air Depot Appears Certain"; *The Macon Telegraph*, June 8, 1941.

"Wellston Gives Way to Warner Robins"; *The Macon Telegraph*, August 16, 1942.

"Wellston May Be Site For Air Depot If Army Engineers Bring It Here"; *The Macon News*, May 19, 1941.

5. *Publications*

"Eighty Base Closures in U.S. Named by McNamara", *Defense Department Digest*, Vol. 1, No. 15, December 15, 1964.

"Formal Dedication Program", Warner Robins Army Air Depot, April 26, 1943.

"McNamara Gives Rationale for New Base Closings", *Defense Department Digest*, Vol. I, No. 13, December 1, 1964.

"The C-141 Starlifter", Walter L. Krane (USAF Hq MAC) December 1968.

"The Pentagon Papers as Published By the New York Times ", (New York: Bantam Books, 1971).

"The U.S. Response to Soviet Military Buildups in Cuba" (Report to the People, John F. Kennedy, October 22, 1962), Department of State Publication 7449, U.S. Government Printing Office 1962.

"U.S. Air Force Gunships: Background Information", USAFOI Pamphlet 68-13, October 1968.

"U.S. Charges of Soviet Military Buildup in Cuba", (Statements of Adlai E. Stevenson, U.S. Representative in the Security Council), Department of State Publication 7458, U.S. Government Printing Office 1962.

"Warner Robins Silver Anniversary Magazine 1953-1968", Anniversary Issue, John R. Adams Publisher

6. *Books*

Abel, Elie, *The Missile Crisis*, (Philadelphia: Lippincott, 1966).

Albertson, Dean, *Eisenhower as President*, (New York: Hill and Wang, 1965).

Berger, Carl, ed., *The United States Air Force in Southeast Asia 1961-1973*, (Washington D.C.: U.S. Government Printing Office, 1977).

Bettinger, Joseph, *Vietnam: A Dragon Embattled*, 2 volumes, (New York: Praeger, 1967).

Blum, John M., et. al., *The National Experience: A History of the United States Since 1865, Part Two*, (New York: Harcourt, Brace & World, 1968).

Chalker, Fussell M., *Pioneer Days Along the Ocmulgee*, (Carrollton, GA. Chalker, 1970).

Dawson, Alan, *55 Days: The Fall of South Vietnam*, (Englewood Cliffs N.J.: Prentice Hall, 1977).

Doyle, Edward, et. al., *The Vietnam Experience: Setting the Stage*, (Boston: Boston Publishing Co., 1981).

Ellsberg, Daniel, *Papers on the War*, (New York: Simon and Schuster, 1972).

Fitzgerald, Frances, *Fire in the Lake*, (Boston: Little, Brown, 1972).

Freidel, Frank, *America in the Twentieth Century*, (Harvard University) (New York: Alfred A. Knopf).

Futtrell, Robert F., *The United States Air Force in Southeast Asia: The Advisory Years to 1965*, (Washington D.C.: U.S. Government Printing Office, 1981).

Galbraith, John Kenneth, *A Life in Our Times* (Boston: Houghton Mifflin, 1981).

Gettleman, Marvin, ed., *Vietnam: History, Documents and Opinions on a Major World Crisis*, (New York: Fawcett, 1965).

Glines, Lt. Col. Carroll V., Jr., *The Compact History of the United States Air Force*, (New York: Hawthorn Books, Inc.).

Goldberg, Alfred, *History of the United States Air Force, 1907-1957*, (Princeton N.J.: D. Van Nostrand Co., Inc.).

Halberstam, David, *The Best and the Brightest*, (New York: Random House, 1969).

Nickson, Bobbe Smith, *A Land So Dedicated*, (Houston County Library Services, 1976).

Holley, Irving Brinton, Jr., *Buying Aircraft: Materiel Procurement for the Army Air Forces* (Washington, D.C.: U. S. Government Printing Office, 1964)

Hoxie, R. Gordon, *Command Decision and the Presidency*, (New York:

Johnson, Lyndon B., *The Vantage Point*, (New York: Holt, 1971).

Kennedy, Robert F., *Thirteen Days: A Memoir of the Cuban Missile Crisis*, (New York, W.W. Norton, 1969).

Kissinger, Henry A., *White House Years*, (Boston: Little, Brown, 1979).

Leckie, George G., *Georgia-A Guide to Its Towns and Countryside*, (Tupper and Love Books, 1954).

Leighton, Richard M. and Cookley, Robert W., *Global Logistics and Strategy, 1940-1943* (Washington D.C.: U. S. Government Printing Office, 1955)

Maclear, Michael, *The Ten Thousand Day War: Vietnam 1945-1975*, (New York: St. Martin.s Press, 1981).

Mason, Herber Molloy, Jr., *A Turbulent History, 1907-1975*, (New York: Mason/Charter).

Matloff, Maurice, *American Military History*, (Washington D.C., 1969).

McCullar, Bernice, *This Is Your Georgia*, (Northport AL: American Southern Publishing Co., 1966).

McNeill, William H., *A World History, Third Edition*, (New York: Oxford University Press 1979).

Ridgway, Matthew B., *The Korean War*, (Garden City NY: Doubleday & Company, Inc., 1967).

Schlesigner, Arthur, *A Thousand Days*, (Boston: Houghton Mifflin).

Shirer, William L., *The Rise and Fall of the Third Reich*, (New York: Simon and Schuster, 1960).

Stevenson, James P., *McDonnell Douglas F-15 Eagle*, (Tallbrook CA: Aero Publishers Inc., 1982).

Stewart, James T., *Airpower—The Decisive Force in Korea*, (Princeton N.J.: D. Van Nostrand Co., 1957).

Taylor, Maxwell, *Swords and Plowshares*, (New York: W. W.. Norton, 1972).

Toland, John, *Infamy*, (New York: Doubleday, 1982).

Westmoreland, William C., *A Soldier Reports*, (New York: Doubleday, 1976).

7. *Persons Interviewed*

Ayres, T. C.
Baldino, Charles A. "Chuck"
Barrs, William D.
Bates, Parks T.
Bauer, Richard F.
Beck, Billy R.
Benefield, Lee J.
Booth, Edna Wellborn Floyd
Bowles, Kenneth
Boyd, Willard L.
Brantley, Louise G.
Braucht, Jack H.
Brown, Debbie
Burns, Zane G.
Cantwell, G. T.
Cater, Thomas J.
Cheshire, Kathy
Clements, Bill
Coefield, Howard H.
Cooper, D. Wayne
Cotton, A. H.
Covington, Carole
Cranshaw, John P.
Crites, Susan W.
Croom, Harold A.
Deep, Johnny
Demarest, Theodore M.
Dixon, Frank
Drake, J. E.
Faulkner, James
Finlayson, Pat
Fussell, Julian R.
Garrett, Thomas L.
Girmsey, John W.
Graham, Henry F.
Grimes, C. B.
Griscom, John H.
Hall, J. Ellsworth, Jr.
Hancock, Betty
Hansen, Lois
Hartman, Vivian P.

Hawkins, Edward A.
Hunt, Thomas
Husband, Herbert
Kinser, Eugene W.
Korson, Skip
Lawson, P. C.
Little, Helen
Loper, James
Loy, MSgt. James M.
Luck, C. G.
Massie, Hannah
McCall, Jill
Miller, TSgt. Jim
Mobley, Clayton
Moore, Cecil
Morris, TSgt. G. M.
Morton, Ilah B.
Munford, Norma M.
Noble, Jeanette
Parris, Robert
Phillips, Hugh
Pierce, Eugene C.
Pogue, Richard
Poole, Thomas F.
Rodriguez, 1st Lt. Kenneth
Ryals, John F.
Sandefur, Larry D.
Sanders, Loraine
Schmitt, Greg
Shaw, Elie
Sheppard, William M.
Silvey, Charles R.
Singleton, Charles R.
Sirmans, Ermal
Stafford, Joseph Sherrill
Tabor, Elizabeth
Taylor, Joyce
Thompson, Harvey
Tucker, Helen Marshall
Walker, Herbert A.
Watson, Clint
Young, William L.

8. *Persons Furnishing Photographs for Consideration*

Adams, John
Beck, Billy R.
Birdsong, John
Bowman, Mary
Braucht, Jack H., Col. USAF (Ret)
Brinson, B. Lewis Jr.
Brown, Butler
Burkhalter, Annelle
Burns, Zane G.
Burwell, Alice
Byrd, Christine
Callahan, Dr. Dan
Cheek, Jesse D.
Clare, Henry
Combs, Charles
Cooper, Robert M.
Cox, W. C.
Crites, Susan W.
Davis, Robert
Day, Albert A.
DeCastro, Mervin A.
Deep, Johnny
Edrington, Lt. Col. P. K.,
Edwards, Billy
Evans, W. E.
Frazer, N. H.
Griscom, John H.
Hails, Robert E., Lt. Gen. USAF (Ret)
Hardman, Allie
Hart, Johnny
Hawkins, Edward A.
Hefley, William T., Lt. Gen. USAF (Ret)
Hemak, TSgt. Edward G.
Holland, Ralph T., Maj. Gen. USAF (Ret)
Howell, J. K.
Hutchings, William D.
Johnson, Capt. James N.
Johnson, Ralph, Col. USAF (Ret)
Keel, Minnie
Kinser, Eugene
Knowles, Stephen M., Lt. Col. USAF (Ret)
Layfield, Mrs. Alva F.
Lingelbach, Lee
Lovik, Orville
Lowe, Helen
Lowery, Margaret
Mantelli, Charles "Chuck"
McDonald, Mrs. W. B.
Meadows, Lois
Monroe, Dilburn
Mueller, Fred
Munford, Norma
Odell, Dave
Oliver, Ann
Pierce, Eugene C.
Sanders, Loraine
Seamans, Robert
Shepherd, Myrtle
Silsbee, E. R., Lt. Col. USAF (Ret)
Steed, J. H., CMSgt. USAF (Ret)
Stevens, Joseph
Sutton, Maj. R. K.
Tabor, Elizabeth
Taylor, Joyce
Westbury, Kenneth
Whipple, Louise
Williams, Martha
Yawn, Leon
Young, Peggy B.

Index

Glossary

A Bomb	Atom Bomb
A&M	Agricultural and Mechanical
A1C	Airman First Class
AACS	Army Airways Communications System
AAF	Army Air Force
AB	Air Base
ADC	Aerospace Defense Command
ADPE	Automated Data Processing Equipment
AFA	Air Force Association
AFAA	Air Force Audit Agency
AFALD	Air Force Acquisition Logistics Division
AFB	Air Force Base
AFCC	Air Force Communications Command
AFCS	Air Force Communications Service
AFLC	Air Force Logistics Command
AFOSI	Air Force Office of Special Investigations
AFPR	Air Force Plant Representative
AFRES	Air Force Reserve
AFSC	Air Force Systems Command
AGE	Aerospace Ground Equipment
ALS	Advanced Logistics System
AMA	Air Materiel Area
AMC	Air Materiel Command
ANG	Air National Guard
ASD	Aeronautical Systems Division
ATC	Air Training Command
ATE	Automatic Test Equipment
AUS	Army of the United States
AUTOVON	Automatic Voice Switching Network
BAMS	Base Automated Mobility System
BC	Before Christ
BMA	Better Management Association
CAD	Computer Aided Design
CAM	Computer Aided Manufacturing
CCP	Consolidation/Containerization Point
CE	Corps of Engineers
CLSS	Combat Logistics Support Squadron
CNC	Computer Numerical Control
COMPS	Contingency Operation/Mobility Planning and Executive System
CONAC	Continental Air Command
D009	Shipment Document Release System
DARTS	Distributor Automated Rapid Transportation System
DC	District of Columbia
DCS	Defense Communications System
DIR	Disassemble, Inspect, and Repair
DLA	Defense Logistics Agency
DO33	Stock Control and Distribution System
DOD	Department of Defense
ECM	Electronic Countermeasures
EDIMGIAFAD	Every Day in Middle Georgia is Air Force Appreciation Day
EEO	Equal Employment Opportunity
EMCS	Energy Monitoring and Control System
FEORP	Federal Employment Opportunity Recruitment Program
FPHA	Federal Public Housing Authority
FY	Fiscal Year
GEEIA	Ground Electronics Engineering-Installation Agency
GED	General Educational Development
GI	Government Issue
GOP	Grand Old Party
GPATS	General Purpose Automatic Testing System
GRIP	Grass Roots Improvement Program
GS	General Schedule
GSA	General Services Administration
H Bomb	Hydrogen Bomb
HEW	Health Education and Welfare
HQ	Headquarters
IM	Item Manager
IRAN	Inspect and Repair as Necessary
JPA	Job Performance Aid
LLLT	Low Light Level Television
LOGAIR	Logistics Airlift
LOGMARS	Logistics Applications of Automated Marking and Reading Symbols
LRC	Logistics Readiness Center
LSS	Logistics Support Squadron
MAC	Military Airlift Command
MACE	Military Airlift Capability Estimator
MAS	Military Airlift Squadron
MATS	Military Air Transport Service
MCAIR	McDonnell Aircraft
MEP	Management Engineering Program
MET	Management Engineering Team
MIA	Missing in Acton
MVP	Manpower Validation Program
NAACP	National Association for the Advancement of Colored People
NATO	North Atlantic Treaty Organization
NC	Numerical Control
NCO	Non Commissioned Officer
NORAD	North American Air Defense Command
OL	Operating Location
OMB	Office of Management and Budget
ORI	Operational Readiness Inspection
PACAF	Pacific Air Force
PDM	Program Depot Maintenance
PHD	Doctor of Philosophy
PMEL	Precision Management and Evaluation Laboratory
PMRT	Program Management Responsibility Transfer
POW	Prisoner of War
PR-MIPR	Purchase Request-Military Interdepartmental Purchase Request
PX	Post Exchange
Prime BEEF	Base Engineer Emergency Forces
QRP	Quick Reaction Package
RAFB	Robins Air Force Base
RAM	Rapid Area Maintenance
RASS	Rapid Air Supply Support
RATS	Rapid Transportation Support
RDF	Rapid Deployment Force
ROTC	Reserve Officer Training Corps
RRC	Robins Resident Center
RVP	Remote Vehicle Processors
Red Horse	Rapid Engineer Deployable Heavy Operations Repair Squadrons, Engineer
SAC	Strategic Air Command
SAM	Surface-to-Air Missile

SAM	School of Aviation Medicine
SBAMA	San Bernardino Air Materiel Area
SC&D	Stock Control and Distribution System
SEA	Southeast Asia
SEAVICS	Southeast Asia Vehicle Inventory Control System
SM	System Manager
SOAP	Spectrographic Oil Analysis Program
SOLE	Society of Logistics Engineers
SRA	Specialized Repair Activity
SSM	System Support Manager
TAALS	Transportation Automated Addressing and Labeling System
TAC	Tactical Air Command
TCTO	Time Compliance Technical Order
TDY	Temporary Duty
TEWS	Tactical Electronic Warfare System
TRC	Technology Repair Center
TSO	Technical Services Occupation
USAF	United States Air Force
USAFE	United States Air Forces in Europe
USSR	Union of Soviet Socialist Republics
V-E	Victory In Europe
V-J	Victory Over Japan
VIP	Very Important Person
WATS	Wide Area Telephone Service
WICS	Warehouse Inventory Control System
WISE	Worldwide Implementation of Supply Economy
WR-ALC	Warner Robins Air Logistics Center
WRAMA	Warner Robins Air Material Area
WRATSC	Warner Robins Air Technical Service Command
WS	Weather Squadron
WSPO	Weapons System Program Office
WWMCCS	Worldwide Military Command and Control System
XR	Directorate of Plans and Programs (Organizational Symbol)

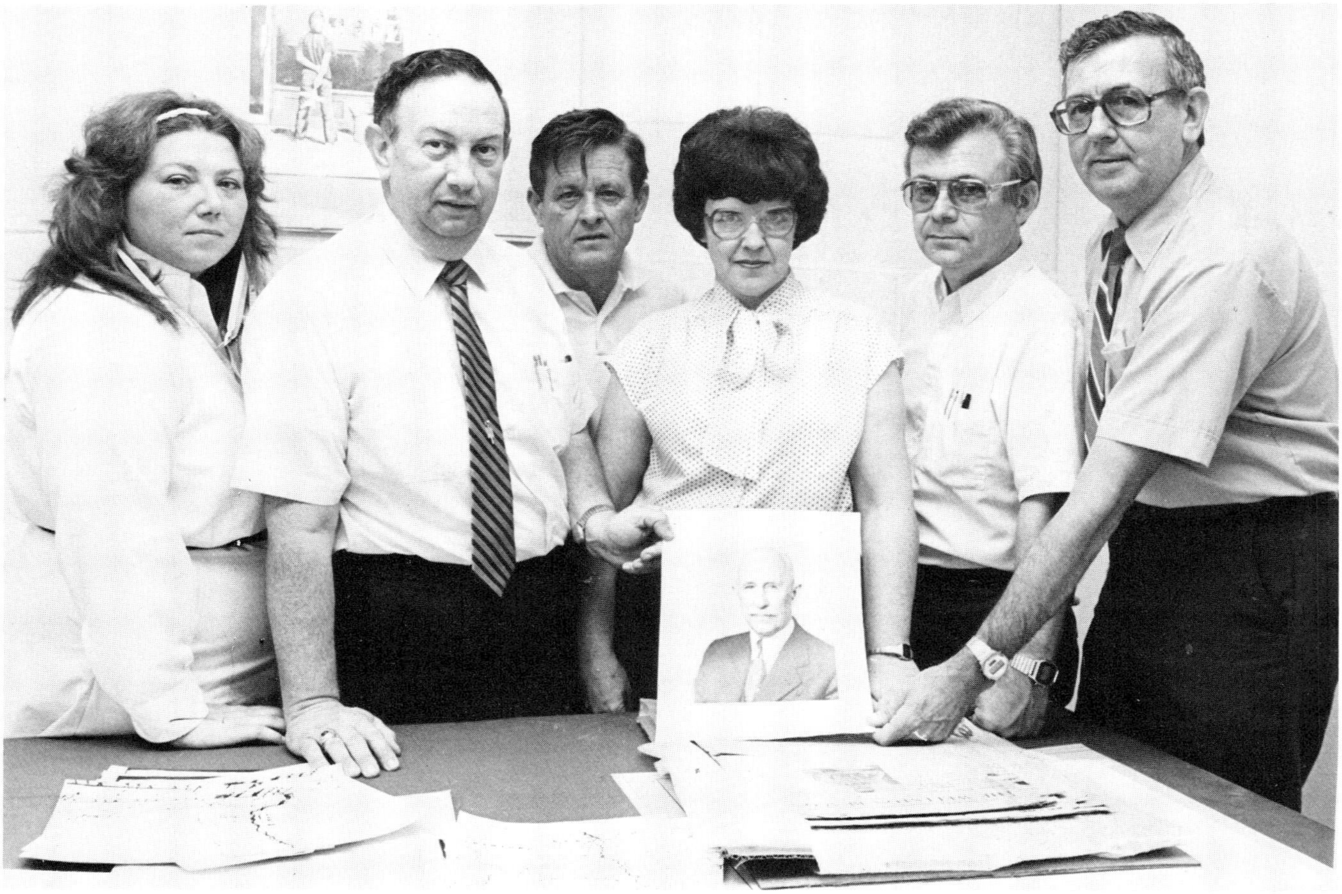

Members of the Pictorial History Team, left to right, *Cara N. Best, B. Lewis Brinson, Jr., Herbert T. Stephens, Carolyn B. Petty, Noble D. Teal and Joseph A. Molony.*

B. Lewis Brinson, Jr., who supervised the Robins AFB history project and guided the team to its successful conclusion, is a native of Emmanuel County, Georgia. He attended the public school system of Emmanuel County but finished his last year of high school at the Georgia Military College at Milledgeville. He has B.B.A. and M.A. degrees from the University of Georgia and an M.P.A. from Harvard. He came to work at Robins AFB in 1958 and is currently Chief of the Plans and Programs Division of the Directorate of Plans and Programs. He and his wife, Judy, have three children and live in Byromville, Georgia.

Carolyn B. Petty, who managed the office where this pictorial history was created, was born in Orlando, Florida. She graduated from the public school system in Warner Robins, Georgia and attended the Robins Resident Center. She came to work at Robins AFB in 1965 and is currently a Program Analyst in the Directorate of Plans and Programs. She and her husband, Johnny, have one son and live in Centerville, Georgia.

Herbert T. Stephens, who wrote the decade of the 1940s, is a native of Montgomery, Alabama. He attended the public school system in that city and has a B.A. degree from Huntingdon College. He came to work at Robins AFB in 1955 and is currently a Program Analyst in the Directorate of Plans and Programs. He and his wife, Mary, have four children and live in Bonaire, Georgia.

Joseph A. Molony, who wrote the decade of the 1950s, is a native of Charleston, South Carolina. He attended the Catholic and public school system of that city and has a B.A. degree from the University of South Carolina. He came to work at Robins AFB in 1961 and is currently the Media Relations Chief in the Office of Public Affairs. He and his wife, Matrice, have four children and live in Warner Robins, Georgia.

Cara N. Best, who wrote the decade of the 1960s, was born to a U. S. Air Force family at Swindon, England. She atttended no less than thirty-five different public schools and has a B.A. degree from Georgia College. She came to work at Robins AFB in 1979 and is currently a Clerical Assistant in the Office of History. She, and her daughter, live in Warner Robins, Georgia.

Noble Dee Teal, who wrote the decade of the 1970s, is a native of Biloxi, Mississippi. He attended the public school system in that city and has a B.S. degree from Mississippi State University. He came to work at Robins AFB in 1959 and is currently a Supervisory Systems Analyst in the Directorate of Distribution. He and his wife, Marilyn, have two children and live in Warner Robins, Georgia.

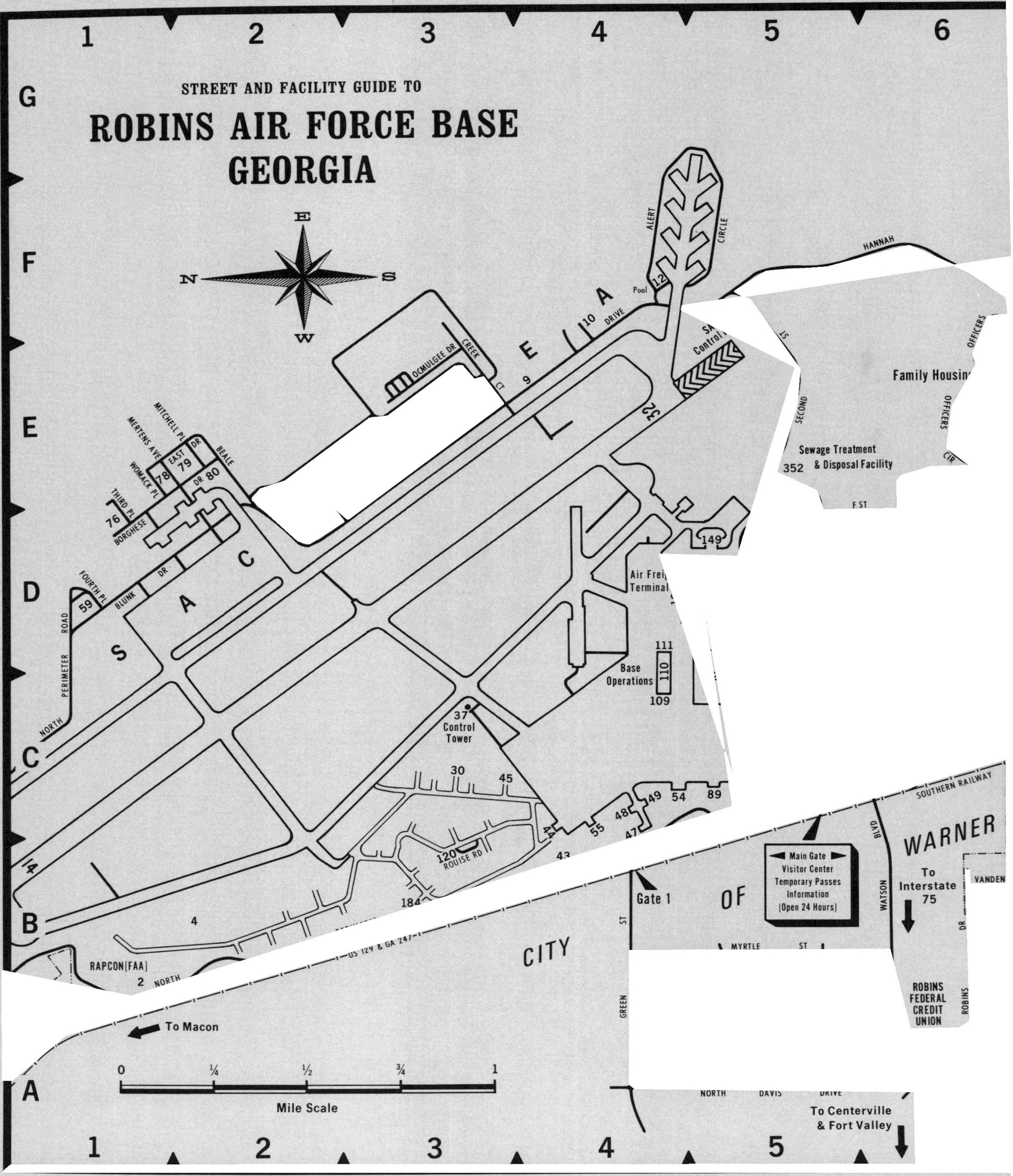

STREET AND FACILITY GUIDE TO
ROBINS AIR FORCE BASE
GEORGIA
E
N
S
W
ALERT
CIRCLE
HANNAH
Pool
12
A
10
DRIVE
E
OCMULGEE DR
CREEK
CT
9
32
Family Housing
OFFICERS
SECOND
Sewage Treatment
& Disposal Facility
352
MITCHELL PL
MERTENS AVE
EAST
DR
BEALE
WOMACK PL
78
79
DR
80
THIRD PL
76
BORGHESE
C
FOURTH PL
59
BLUNK
DR
A
S
PERIMETER
ROAD
NORTH
149
Air Freig
Terminal
111
110
109
Base
Operations
37
Control
Tower
30
45
49
54
89
48
55
47
44
43
120
ROUISE RD
184
4
RAPCON(FAA)
2
NORTH
US 129 & GA 247
CITY
OF
WARNER
SOUTHERN RAILWAY
Main Gate
Visitor Center
Temporary Passes
Information
(Open 24 Hours)
Gate 1
GREEN
ST
MYRTLE
ST
BLVD
WATSON
To
Interstate
75
VANDEN
DR
ROBINS
FEDERAL
CREDIT
UNION
ROBINS
To Macon
0
¼
½
¾
1
Mile Scale
NORTH
DAVIS
DRIVE
To Centerville
& Fort Valley
1
2
3
4
5
6
A
B
C
D
E
F
G